PHYSICS OF MATTER

Ognjen Grkinic

TAP

Toronto Academic Press

PHYSICS OF MATTER

Ognjen Grkinic

Toronto Academic Press
4164 Lakeshore Road
Burlington ON L7L 1A4
Canada
www.tap-books.com

Email: orders@arclereducation.com

ISBN: 978-1-77956-343-9

ABOUT THE AUTHOR

Ognjen Grkinić is a physicist with a profound expertise in applied and computer physics, particularly in data science and plasma physics. He is currently pursuing his PhD at the Faculty of Physics, University of Belgrade, where his research focuses on the behavior of ionized gases and plasma within the Attosecond-Plasmas Lab. Ognjen's academic achievements include a Master's degree in applied and computer physics, where he developed advanced software for processing scientific data, earning a near-perfect GPA of 9.80/10.00. In addition to his research, Ognjen has contributed significantly to education, having taught physics and mathematics at both high school and primary school levels. His passion for teaching is matched by his meticulous approach to scientific inquiry, which is evident in the books and papers he has authored and edited in the field of physics and other natural sciences. Beyond academia, Ognjen Grkinić's diverse experiences in research, education, and professional work underscore his commitment to advancing the field of physics and contributing to the broader scientific community.

Table of Contents

List of Figures

List of Table

PREFACE

Physics of Matter explores the fundamental principles governing the behavior and properties of materials at atomic and molecular scales, encompassing a wide range of topics from the structure of atoms and molecules to the collective behavior of solids, liquids, and gases. It delves into the microscopic interactions and macroscopic phenomena that dictate the physical properties of matter, such as electrical conductivity, thermal expansion, and magnetic behavior. By studying the arrangement and dynamics of particles within different states of matter, Physics of Matter provides a foundational understanding essential for advancements in areas like materials science, nanotechnology, and condensed matter physics. This discipline bridges theoretical concepts with practical applications, aiming to elucidate the underlying physics that governs the diverse behaviors and functionalities exhibited by various substances in nature and engineered systems.

This book, 'Physics of Matter,' is structured into eight comprehensive chapters, each aimed at unraveling the fundamental principles that govern the behavior and properties of matter at different scales. Chapter 1 serves as an introduction to the foundational concepts of physics relevant to matter. Chapter 2 delves into the structure and properties of atoms, which are the building blocks of all matter. It explores the intricacies of atomic structure and introduces the principles of quantum mechanics that govern atomic behavior.

In Chapter 3, the focus shifts to molecular structure and bonding. This chapter explores how atoms combine to form molecules and the types of bonds that hold them together. Chapter 4 introduces statistical physics, which provides a framework for understanding the collective behavior of particles in systems ranging from gases to complex fluids.

Chapter 5 focuses on solids, detailing their crystalline structures, mechanical properties, and behaviors under different conditions. Chapter 6 shifts to liquids and gases, examining their fluid dynamics, phase transitions, and thermodynamic properties.

Chapter 7 broadens the scope to soft matter and complex fluids, which encompass a diverse range of materials with unique properties and behaviors. Finally, Chapter 8 discusses modern developments in the physics of matter, showcasing recent advancements and applications in areas like nanotechnology, advanced materials, and interdisciplinary research.

This book is designed for students, researchers, and professionals interested in gaining a comprehensive understanding of the principles governing matter physics. By bridging theoretical insights with practical applications across these diverse chapters, we aim to provide a valuable resource that fosters deeper understanding and innovation in fields as varied as materials science, physics, chemistry, and engineering."

—Author

CHAPTER

1 Fundamentals of Physics of Matter

LEARNING OBJECTIVES

After studying this chapter, you will be able to:

1. Understand the fundamental properties of electrons and their role in atomic structure;
2. Describe the composition and interactions of atomic nuclei;
3. Explain Schrödinger's equation and its significance in quantum mechanics;
4. Identify the timescales and frequencies associated with electronic motions;
5. Analyze the impact of electric and magnetic fields on spectral lines;
6. Apply perturbation theory and variational methods to atomic systems;
7. Discuss the principles and applications of advanced spectroscopy techniques;
8. Explore the applications of atomic physics in modern technologies such as atomic clocks and quantum computing.

KEY TERMS FROM THIS CHAPTER

Doppler effect

Nuclear composition

Spectral broadening

Zeeman effect

Electron properties

Schrödinger's equation

Stark effect

UNIT INTRODUCTION

This chapter examines matter based on atomic and quantum laws from contemporary physics, focusing on the fundamental aspects of atomic physics. It explores the basic constituents, interactions, and uses of atomic physics, revealing the principles that direct the behavior of matter, from particles to broader applications in technology and astrophysics.

The chapter begins with an overview of the fundamental components of matter: electrons, protons, and neutrons. Electrons, with their charge, mass, and spin, are distributed in orbitals around the nuclei, made up of protons and neutrons. Understanding these fundamental particles and their interactions is crucial to comprehending atomic and molecular structures. Schrödinger's equation provides a mathematical framework to understand the wave-like behavior of particles and offers insights into the probability distributions of electrons within atoms (Landau & Lifshitz, 1980).

The interaction of energy among elementary charges is key to understanding atomic equilibrium and bonding. The speed and frequencies of electronic movements highlight the wave dynamics of atomic behavior, demonstrated through the fine-structure constant and electron velocities.

This chapter also explores the structure and arrangement of matter, examining the spatial dimensions of electron orbits and the frequencies that cause their oscillations. It places atomic behavior within a broader context, providing a comprehensive understanding of physical phenomena (Rajagopal & Wilczek, 2001).

The chapter then focuses on the examination of spectra and the effects of broadening, crucial aspects of atomic physics. Spectroscopy techniques enable the accurate study of atomic emissions and absorptions, providing information about energy levels, transitions, and fundamental quantum states. The chapter explores the mechanisms of range broadening, including Doppler broadening from thermal motion and collisional forces that affect atomic spectra in different conditions (Andelman, 2006).

Advanced quantum concepts are presented to deepen understanding of atomic systems. Perturbation theory explains the behavior of atomic states under small changes in potential energy, while variational methods provide approximate solutions to the Schrödinger equation, essential for calculating atomic energies accurately. The fine structure of atomic spectra is explained by studying spin-orbit coupling, revealing the relationship between spin and orbital angular momentum (Einstein, 1936).

The practical applications of atomic physics are emphasized, discussing experiments like spectral analysis for atomic evaluation and quantum interference and coherence effects. These methods connect theoretical knowledge with real-world applications, advancing fields such as precise timekeeping in atomic clocks and the development of quantum computing technology (Halbritter et al., 1995).

Overall, this chapter provides a comprehensive overview of atomic physics, exploring the core principles that govern matter at the atomic level and showcasing the wide range of applications and implications of atomic physics in today's technological and scientific fields. Understanding these core principles and their real-world implications enhances our comprehension of the cosmos (Naruse et al., 2013).

Remember

Remember that the fundamentals of the physics of matter encompass understanding the properties and interactions of electrons, nuclei, and their microscopic behaviors, crucial for explaining the macroscopic properties of materials.

1.1. BASIC INGREDIENTS

Over time, a significant amount of experimental evidence was gathered by physicists along with chemists within the latter part of the 19th and the start of the 20th century. This evidence ultimately led to the consensus that all forms of matter, which could be helium, metallic screw, solid ice, a block of wood, a mobile phone, or even a bee, are made up of countless electrons along with atomic nuclei (Lavelle et al., 2017).

1.1.1. Electron Properties

The electron is a remarkable elementary particle with a mass of approximately $\left(m_e \cong 9.11\times10^{-31}\right)$ along with a negative charge $\left(q_e \cong -1.60\times10^{-19}\right)C$ (Castro et al., 2009).

1.1.2. Nuclear Composition

The inner structure of nuclei is quite complicated, via length scales upon the order of $\left(\cong 10^{-15}\right)$ m along with excitation energy around $\left(\cong 10^{-13}\right)J$. Nuclear features have little significance when it comes to the everyday properties of matter. In practical terms, matter can be described by considering nuclei just like point-like particles without any significant structure (Haensel & Zdunik, 2003).

The nucleus consists of (Z) protons along with (A) nucleons (protons & neutrons), which have a positive charge of (Zq_e) or a mass of approximately $\left(M \cong A\times1.66\times10^{-27}\right)$ kilogram (atomic mass units).

1.1.3. Interactions Among Components

When one ignores relativistic effects along with the relationship of matter using its surroundings, the inner microscopic relations within its components are solely electromagnetic. The movement of both nuclei and electrons within the sample is determined by the energy operator known as the Hamiltonian:

$$H_{\text{tot}} = T_n + T_e + V_{ne} + V_{nn} + V_{ee} \qquad (1.1)$$

where; (T_n) is the kinetic energy of the nuclei; (T_e) is the kinetic energy of the electrons; (V_{ne}) is the potential energy describing the attraction between nuclei and electrons; (V_{nn}) is the describes the nucleus-nucleus repulsion; and (V_{ee}) is the electron-electron repulsion.

$$T_n = \frac{1}{2}\sum_{á} \frac{P_{á}^2}{M_{á}} \quad (1.2)$$

This equation indicates the kinetic energy that exists in nuclei (T_n), the $(\sum_\alpha)$ summation symbol denotes the total energy across all nuclei, while $(\alpha),(P_\alpha)$ stands for the conjugate momentum for the position (R_α), and (M_α) stands for the mass from the nucleus (α) (Kilduff et al., 1996).

The energy related to the movement of electrons:

$$T_e = \frac{1}{2m_e}\sum_i P_{ri}^2 \quad (1.3)$$

where; (T_e) denotes the kinetic energy given by electrons. The conjugate momentum for r_i is denoted as P_{r_i}. The mass for each electron is represented by m_e. The summation across every electron is indicated by (i).

Exploring the potential energy resulting from the interaction between nuclei and electrons:

$$V_{ne} = -\frac{q_e^2}{4\pi\grave{o}_0}\sum_\alpha\sum_i \frac{Z_\alpha}{|R_\alpha - r_i|} \quad (1.4)$$

The symbol (V_{ne}) represents the potential energy resulting from the attractive force within the nuclei (α) along with the electrons (i). In this equation, (q_e) represents the fundamental charge, $(\grave{o}_0)$ represents the permittivity for free space, $(Z_{á})$ represents the charge number regarding the nucleus (α), (R_α) represents the position for the nucleus (α), or (r_i) represents the position for the electron (i) (Zhao et al., 2018).

The potential energy that characterizes the repulsion between two nuclei:

$$V_{nn} = \frac{q_e^2}{4\pi\grave{o}_0}\sum_{\alpha\neq\beta} \frac{Z_\alpha Z_\beta}{|R_\alpha - R_\beta|} \quad (1.5)$$

(V_{nn}) represents the potential energy resulting from the repulsion across the nuclei (α) or (β). The charged number for the nuclei (α) and (β) are denoted by (Z_α) and (Z_β), respectively, while their locations are represented by (R_α) and (R_β).

The potential energy that characterizes the repulsion between electrons:

$$V_{ee} = \frac{q_e^2}{4\pi\grave{o}_0}\sum_{i\neq j} \frac{1}{|r_i - r_j|} \quad (1.6)$$

The symbol (V_{ee}) denotes the potential energy arising from the repulsive interaction between electrons (i) and (j), with (r_i), and (r_j) representing the locations of electrons (i) and (j), respectively.

These equations describe the interaction between nuclei and electrons within a system, governed by electromagnetic forces, under the assumption of non-relativistic conditions and ignoring interactions with the environment (Wang et al., 1999).

1.1.4. Schrödinger's Equation and Microscopic Interactions

The evolution from a state $ket\left(\left|\hat{\imath}\left(t\right)\right.\right)$ that encompasses all the quantum-mechanical data about the movement of electrons and nuclei is governed by Schrödinger's equation.

$$i\hbar\frac{d}{dt}\left|\hat{\imath}\left(t\right)\right. = \widehat{H_{\text{tot}}}\left|\hat{\imath}\left(t\right)\right. \tag{1.7}$$

This equation, derived from Hamiltonian (1.1), appears to be straightforward and applicable in a wide range of scenarios. The simplicity and universality of this concept suggest that it is theoretically feasible to comprehend the observed actions of every isolated microscopic object by studying its microscopic relations. Unfortunately, only a limited number of simple and imagined cases have precise answers for Eqn. (1.7). When trying to find an approximate mathematical solution for Eqn. (1.7), one quickly encounters difficulty. The amount of information in an N-particle ket grows exponentially with N, surpassing the capabilities of a single computer (Lorin et al., 2007).

To understand the properties of simple materials like gases and solids, it is important to use complex estimates to solve the equations. Understanding observed characteristics and predicting new properties for material systems relies on the use of intelligent approximations to Eqn. (1.7). These estimates are crucial for understanding how the macroscopic features of matter are connected through the microscopic relations (Bernard et al., 2011).

Applying approximations to Eqn. (1.7) to understand observed properties and predict new features of material systems requires advanced skills. These approximations are essential tools that link macroscopic properties of matter to the underlying microscopic interactions. This textbook provides a thorough exploration of various phenomena in matter physics, offering fundamental conceptual tools for deeper understanding. The suggested methods are simplified models that serve as a starting point for grasping complex phenomena (Chen et al., 2017).

1.2. TYPICAL SCALES

The dimensional Scales within Hamiltonian: (H_{tot}).

The Hamiltonian (H_{tot}) describes the dynamics of a system and incorporates various dimensional scales derived from physical constants. For example, the speed of light (c) plays a significant role in the Hamiltonian. Notably, in the total Hamiltonian, the elementary charge (q_e) and the electromagnetic constant (ϵ_0) consistently appear together as a fixed pair (Shen et al., 2015).

$$e^2 \equiv \frac{q_e^2}{4\pi\epsilon_0} = 2.3071 \times 10^{-28} \text{ Jm}$$

The dimensions are given by the product of energy × length. The distinctive length is obtained by combining the values for (e^2), Planck's constant $(\hbar)$, along with a mass of electron (m_e).

$$a_0 = \frac{\hbar^2}{m_e e^2} = 0.529177 \times 10^{-10} \text{ m} \quad (1.8)$$

The Bohr radius represents the standard scale for electrical motions in atoms. Most small structures and patterns in matter naturally have spacings on the order of the Bohr radius (a_0). For example, microscopes can be used to investigate atomic positions (Sullivan & Artino, 2013)

Atomic Force Microscopes (AFMs) use a precise tip to scan solid surfaces, measuring the forces between surface atoms and the tip. Scanning tunneling microscopes (STMs) measure the electronic tunneling current between the tip and the surface while keeping a very small distance between them. These techniques reveal that atoms in materials are typically spaced only a fraction of a nanometer apart, generally between 2 and 10 angstroms (Anazawa et al., 2004) (Figure 1.1).

- **Interaction Energy of Elementary Point Charges:** The energy that arises from the interaction between two elementary charge points at a typical distance (a_0) can be expressed as:

$$E_{Ha} = \frac{e^2}{a_0} = \frac{m_e e^4}{\hbar^2} = 4.35975 \times 10^{-18} \text{ J} = 27.2114 \text{ eV} \quad (1.9)$$

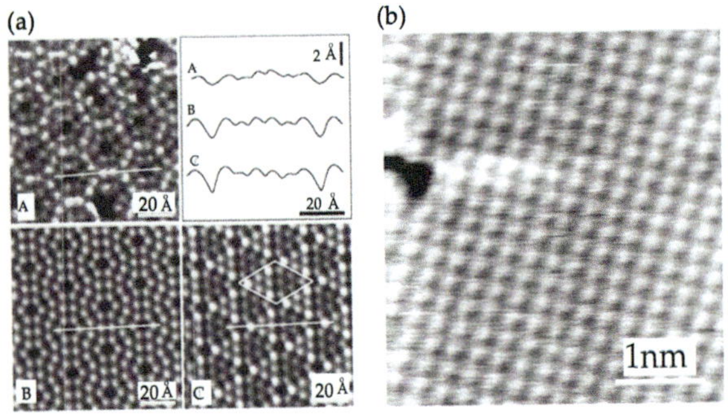

Figure 1.1. *The AFM (A) and STM (B, C) topography images of a clean silicon (111) surface. The STM map visualizes empty (B) and filled (C) electronic states.*

Source: R. Erlandsson, Creative Commons License.

The Hartree energy, also known as the natural energy scale, is a crucial factor in understanding the behavior of a single electron in atomic systems. Practically, the electron Volt (eV) is a widely used unit of energy, with (1 eV = 0.0367493 E_h). Nuclear charge aspects $\left(Z\alpha \approx 10^2\right)$ can significantly increase the electron-nucleus connecting constant $\left(e^2\right)$, By decreasing the distance $\left(10^{-2}\right)$, between electrons and nuclei, the binding energies for electrons can be significantly increased, resulting in a substantial boost of upwards of four orders of magnitude $\left(\left(10^4\right)E_h \approx 272\,keV\right)$. On the other hand, there are times when precise equilibriums can result in electrical excitation energy as low as 1 meV. The movements of the nuclei typically involve lower energies $10^{-4} \div 10^{-3}$ E_h) and velocities due to their much greater mass compared to other factors in Eqn. 1.2.

1.2.1. Timescale and Frequency of Electronic Motions

The relationship between the timescale that electronic motions and its energy scale were inverted:

$$t_0 = \frac{\hbar}{E_{\mathrm{Ha}}} = \frac{3m_e^4}{\hbar} = 2.41888 \times 10^{-17}\ \mathrm{s.} \quad (1.10)$$

Oscillations among a period for $\left(2\pi t_0\right)$ posessing *frequency* $\left(\nu_0 = \omega_0/(2\pi) = E_{\mathrm{Ha}}/(2\pi\hbar) = 6.5797 \times 10^{15}\ \mathrm{Hz}\right)$

1.2.2. Electron Velocity and Fine-Structure Constant

The velocity of electrons can be based on a specific ratio:

$$v_0 = \frac{a_0}{t_0} = \frac{E_{\mathrm{Ha}}}{m_e} = \frac{e^2}{\hbar} = 2.18769 \times 10^6\ \mathrm{m/s}, \quad (1.11)$$

which is approximately 1% of the speed of light, providing a retrospective justification for the initial disregard of relativity. More specifically, a ratio:

$$\alpha = \frac{v_0}{c} = \frac{E_{\mathrm{Ha}}}{m_e c^2} = \frac{e^2}{\hbar c} = 7.29735 \times 10^{-3} = \frac{1}{137.036}, \quad (1.12)$$

This fine-structure constant, which quantifies the significance of relativistic changes for the nonrelativistic movement of electrons, is well-known in the field. Heavier nuclei generally have slower movement compared to electrons, which means that relativistic changes in their motion are usually not significant (Bekenstein, 2002).

1.2.3. Comparison of Electron Length and Frequency Scales

Figure 1.2 illustrates a contrast between the size and speed of electrons within matter and the wavelengths along with frequencies for electromagnetic waves utilized to study matter. The radiation's wavelengths align with the standard interatomic distances of approximately $(\sim 10^{-10})$ m within the X-ray region. On the other hand, the frequencies of radiation (and therefore the energies of photons) align with the standard atomic frequency scale (ν_0) (and the corresponding energy scale E_h) within the ultraviolet (UV) region. In this particular range of the electromagnetic spectrum, the wavelength is significantly larger compared to the size $(\sim a_0)$ for an atom. It is about a third of magnitude bigger than the average atomic size (Brus, 1984).

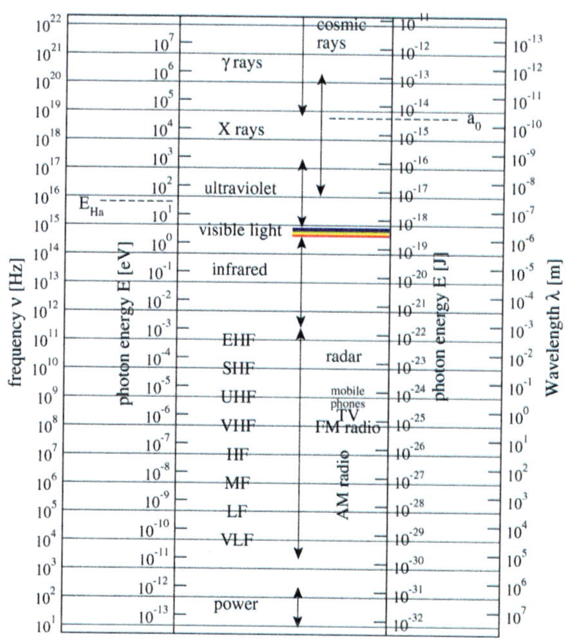

Figure 1.2. *The electromagnetic radiation spectrum is defined by its frequency (v), photon energy (E = kω = 2πkv), and wavelength (λ = c/v). Observe the hues of visible light as the distinctive atomic measurements a0, as stated in Eqn. (1.8), and EHa, as stated in Eqn. (1.9).*

Source: Nicola Manini, Creative Commons License.

The wavelengths and energies commonly observed in the motion for nuclei in matter fall within the infrared (IR) region within the electromagnetic spectrum, typically around $(\sim 5)\pm$ or $\sim 10\pm$ (Popescu et al., 2017).

1.2.4. Structure of Matter

In a matter, electrons are drawn towards the most powerful charges, which constitute nuclei. This attraction is explained by the Eqn. (1.4). To avoid electrons collapsing within nuclei, their kinetic energy, described within Eqn. (1.3), along with Heisenberg's uncertainty connection comes into play. Instead, atoms are formed with a finite size, which is approximately (a_0). (Chabardes et al., 2003).

The behavior of matter on a larger scale, including its elasticity, heat transportation, along with heat capacity, may be accurately explained by studying the movements of atoms or groups of atoms. The movements and interactions for atoms are dictated by the intricate movement of electrons with charges along with nuclei, as outlined in Eqns (1.1) and (1.7), usually examined using the adiabatic division scheme. Grasping the intricacies of a limited group of electrons interacting with multiple nuclei, along with the movement of these nuclei, forms the basis for comprehending interatomic bonding. This fundamental knowledge is what underlies the existence of particles and reduced states of matter (Huhtamo, 2008). The techniques and estimations developed for systems with both small numbers of atoms and large groups of particles lead to new insights and phenomena related to larger-scale objects. Understanding matter at various scales is crucial for advancing fields such as chemistry, materials science, and condensed matter physics (Castleman & Bowen, 1996).

1.3. SPECTRA AND BROADENING

Since the end of the 19th century, scientists have utilized a range of methods to study the complex phenomena of matter's excitations. A significant amount of proof has been gathered using two main types of spectroscopies: absorption and emission.

1. **Absorption:** During absorption spectroscopy, a focused stream of light of a single wavelength (which may or may not be visible) is passed through the material. When the frequency of light aligns with a specific transition within the sample, the sample absorbs photons at this frequency, causing electronic or vibrational transitions (Konjević, 1999). The energy extracted from the initial beam is commonly emitted in arbitrary directions, resulting in a decrease in the beam's intensity shown as a function of its light frequencies (or wavelength), as depicted in Figure 1.3.

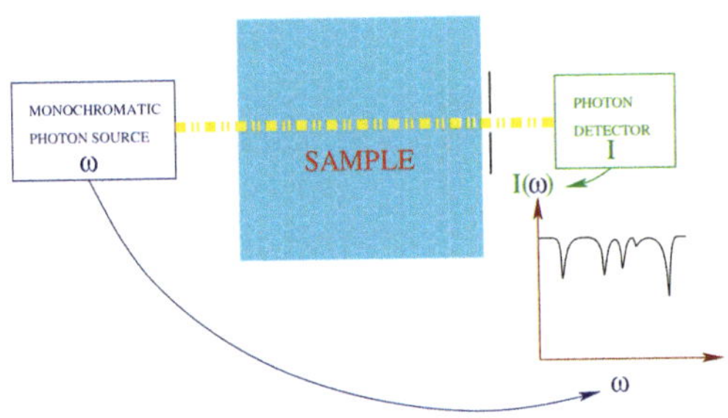

Figure 1.3. *An illustration of the arrangement for absorption spectroscopy.*

Source: Nicola Manini, Creative Commons License.

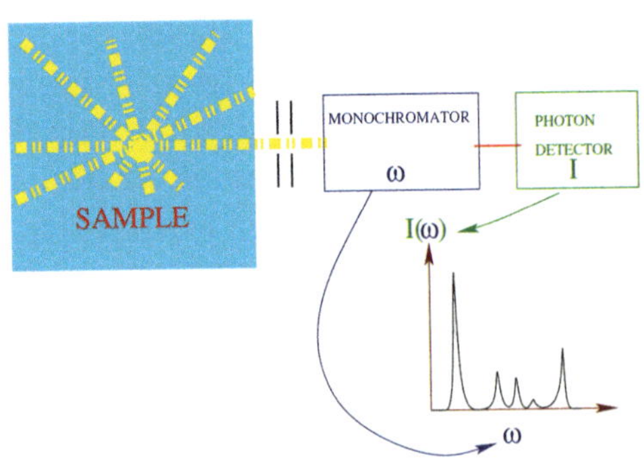

Figure 1.4. *A visual representation of the arrangement for emission spectroscopy.*

Source: Nicola Manini, Creative Commons License.

2. **Emission:** In emissions spectroscopy, the specimen undergoes excitation to higher power states, typically achieved by heating about a flame or bombarding with high-energy electrons. The light released by the sample after its deexcitation shifts is measured and the magnitude is recorded based on frequency (Figure 1.4).

Researchers frequently collect spectra of both kinds across a wide range of wavelengths, including visible, infrared (IR), ultraviolet (UV), and X-ray, to study the characteristics of various samples, including gaseous, liquid, or solid substances. The examination of colors observed by the human eye when objects are illuminated using white light may be seen as a basic type of emission spectroscopy experiment (Hindmarsh et al., 1967).

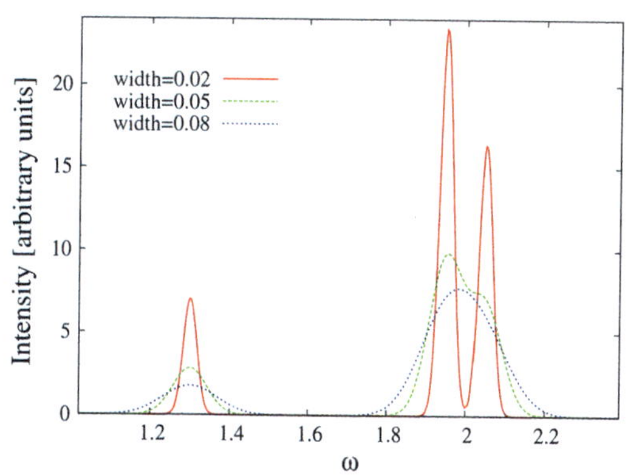

Figure 1.5. *Increasing the width of a spectrum restricts the level of detail that can be observed. If the width of the lines within the spectrum is greater than the distance between two adjacent lines (as shown by the dotted curve), the lines appear as a single bump. This causes the fine details of the two nearby lines to be indistinguishable.*

Source: Neil Ashby, Creative Commons License.

Atomic along with molecular spectra frequently display distinct, narrow peaks called "lines" that represent transitions among quantum states ($|i\rangle$) while ($|f\rangle$). These transitions are defined from discrete energy levels (E_i) and (E_f). The angle of incidence of such a distinct peak is ($\omega_{if} = |E_i - E_f|/\hbar$).

In real-world situations, absorption and emission peaks are never completely sharp. The ability to resolve spectral descriptions is limited by broadening effects, as shown in Figure 1.5. Expanding outcomes arise from the collective influences of:

(i) The spectrometer has a limited testing resolution.

(ii) Broadening caused by the limited duration of excited states.

(iii) Doppler broadening (Myers, 1998).

Experimental Resolution: Constrained by the capabilities that the monochromator, potential noise within photon detectors along with electronics, deviations within sample homogeneities, or fluctuations in external fields. Resolution broadening often leads to a Gaussian line form that can be minimized through technical improvements.

The elementary Schrödinger concept predicts that atomic eigenstates remain stationary because it does not consider interactions with the constantly fluctuating ray's field. The ground state stays completely still, while excited states naturally transition to lower-energy eigenstates because of this connection (Patsourakos & Klimchuk, 2006).

A single atom in an excited state undergoes spontaneous decay at an indeterminate and unknown point in time. Yet, a substantial collection of atoms that are initially stimulated suffers a predictable decay over a period (t). This decay follows a specific pattern, where the mean number of atoms that remain to be decayed is determined. This pattern is known as the decay law, and it establishes the *lifetime* τ of a single atom and its rate of decay $\gamma = 1/\tau$ for that atomic level (Somsen et al., 1996).

$$\left([N](t) = N_0 e^{-\gamma t} = N_0 e^{-t/\tau}\right)(Eq.1.13).$$

1.3.1. Decay Rate and Spectral Broadening

The deterioration rate (γ) is constant as time passes, implying that the probability of an atom decaying does not alter while it remains within its excited state. This attribute defines the process of decay that occurs randomly and uniformly. Atoms generally do not stay within an excited state for more than a few multiples of their standard lifetime (τ). Therefore, (τ) determines the average length of a spectroscopic experiment that includes these excited states. According to the time-energy uncertainties principle, the energy on an atomic level will not be determined more accurately than:

$$(\Delta E \approx \hbar\gamma)(Eq.1.14).$$

The limited duration of atomic levels results in the widening of spectral lines, that might not have an infinite degree of precision. The phenomenon of lifetime-induced broadening is observed in the spectrum within the Lorentzian peak profile.

$$I(\omega) = I_0 \frac{\gamma^2}{(\omega - \omega_{if})^2 + \gamma^2} \quad \text{(Eq. 1.15)}$$

Atomic states with high levels of excitement usually have lifetimes in the range of a few nanoseconds causing a small natural expansion (γ) of the spectrum in the form of microelectronvolts (μeV) (Pericoli et al., 1994).

1.3.2. Doppler Broadening

Random thermal movement of particles within a gas-phase sample causes line broadening, which is referred to as the Doppler effect. While molecules or atoms are in motion relative to the detector, their noticed transition frequency can either be shifted towards the blue or red end of the spectrum depending on their direction of movement (Biraben et al., 1974) (Figure 1.6).

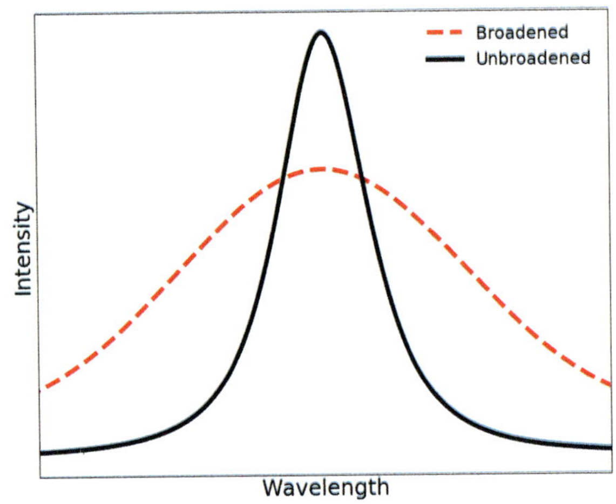

Figure 1.6. *An example of a Doppler broadened line profile. The solid line represents an un-broadened emission profile, and the dashed line represents a broadened emission profile.*

Source: Planet Physics Guy, Creative Commons License.

Considering the non-relativistic character of heat velocities, the description of this Doppler shift in frequency is quite simple:

$$\omega = \omega_{if}\left(1 \pm \frac{v_x}{c}\right) \quad \text{(Eq. 1.16)}$$

The distribution of molecule velocities is determined by the gas temperature T. The number of molecules via velocity component vx along the path of detection can be calculated as an average.

$$\frac{dn(v_x)}{dv_x} = \frac{N}{\sqrt{2\pi k_B T M}} \exp\left(-\frac{M v_x^2}{2k_B T}\right) \quad \text{(Eq. 1.17)}$$

In this case, (v_x) reflects the molecular speed component together with the path of detection direction (Gianfrani, 2016). This implies a Maxwell-Boltzmann distribution that depends on the gas's temperature (T). The radiation intensity then propagates around the remaining frequency (ω_{if}) by:

$$I(\omega) = I_0 \exp\left(-\frac{Mc^2}{2k_B T} \frac{(\omega - \omega_{if})^2}{\omega_{if}^2}\right) \quad \text{(Eq. 1.18)}$$

This widening effect results in a total width at half of the maximum value:

$$\Delta\omega_{Doppler} = \omega_{if} \sqrt{\frac{8 \ln(2)\, k_B T}{Mc^2}} \quad \text{(Eq. 1.19)}$$

Suppose $\lambda = 2\pi c/\omega$, the expansion of the range in terms of wavelength is equal to the frequency:

$$\frac{\Delta\lambda_{\text{Doppler}}}{\lambda_{if}} = \sqrt{\frac{8 \ln(2)\, k_B T}{Mc^2}} \quad \text{(Eq. 1.20)}$$

Due to the inverted square root dependency $((M^{(\frac{-1}{2})}))$ at Eqns. (1.19) and (1.20), molecules or atoms with lower mass and higher velocity at a particular temperature are more significantly affected by Doppler broadening. For instance, the atomic hydrogen spectrum exhibits a noticeable phenomenon known as Doppler broadening. In this case, the Hα line within a hydrogen gas with a temperature of 300K has a width of approximately $(\Delta\omega_{Doppler} \approx 10\,\mu\text{eV})$ (Yesilyurt et al., 2012).

Based on statistical rules, the various factors that contribute to broadening come together to produce a combined broadening effect $\Delta\omega_{\text{tot}} = \sqrt{(\Delta\omega_{\text{exp}})^2 + \gamma^2 + (\Delta\omega_{\text{Doppler}})^2}$.

1.3.3. Pressure Broadening and Collisional Effects

Pressure broadening, also referred to as collisional broadening, is a phenomenon resulting from the interactions among molecules or atoms in a gas, particularly at elevated pressures. These interactions impact the energy levels within atoms or molecules, resulting in a widening of spectral lines. One can grasp the concept of this phenomenon by considering how collisions affect the energy levels of the particles involved in emission or absorption (Hindmarsh & Farr, 1973).

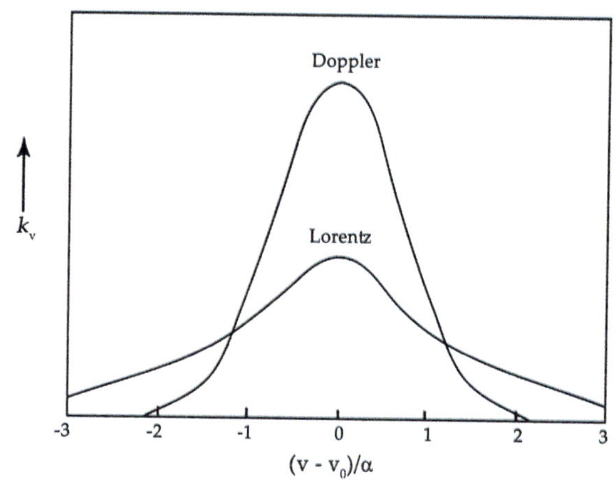

Figure 1.7. *Demonstration of Lorentz and Doppler line shapes for the same intensities and line widths.*

Source: Robert L. Cook, Creative Commons License.

1.3.3.1. Lorentzian Broadening Due to Collisions

The spectral line form that arises through collisional broadening is commonly characterized as Lorentzian. The spectrum line's width is directly proportional to the frequency of collisions, which escalates with pressure (Claveau et al., 2001). The FWHM of a Lorentzian line shape may be mathematically represented:

$$\Delta\nu_{coll} = \frac{1}{2\pi\tau}$$

whereas (τ) The mean time across collisions is the average duration between two consecutive collisions. The correlation between pressure (P) The collision speed can be calculated by applying the laws of ideal gas along with kinetic theory. Considering a gas at a certain temperature (T) along with pressure (P), the rate of collision (ν_{coll}) might be estimated as:

$$\nu_{coll} \approx n\sigma v_{th}$$

In which (n) is the density of the gas, \(σ)shows collision cross-section, whereas (v_{th}) The thermal velocity refers to the average speed at which particles move due to their thermal energy. Applying the ideal gas laws $\left(n = \frac{P}{k_B T}\right)$, in which (k_B) represents the Boltzmann constant, one gets:

$$\Delta\nu_{coll} \propto P$$

Therefore, the widening of any spectral line correlates directly with the pressure (Danos & Geschwind, 1953).

1.3.3.2. Impact Parameter and Phase Shift

When particles collide, their impact parameter (b) along with the phase shift (ϕ) can be used to explain the effect created by the collision. When considering the interaction within particles during a distance (r), the phase shift (ϕ) might be determined via the potential $(V(r))$.

$$\phi = \int_{r_{min}}^{\infty} \frac{V(r)}{v}\, dr$$

where; (r_{min}) represents the distance for the closest approach, while v denotes the relative velocity between the particles (Li et al., 2017).

1.3.4. Stark Effect and Electric Field Influence on Spectra

The Stark effect is observed when atomic and molecular spectral lines break or shift due to the influence of an electric field from the outside. This phenomenon is like the Zeeman effect, in which a magnetic field has an impact on the spectral lines. Two types constitute the Stark effect: the linear Stark impact along the quadratic Stark impact.

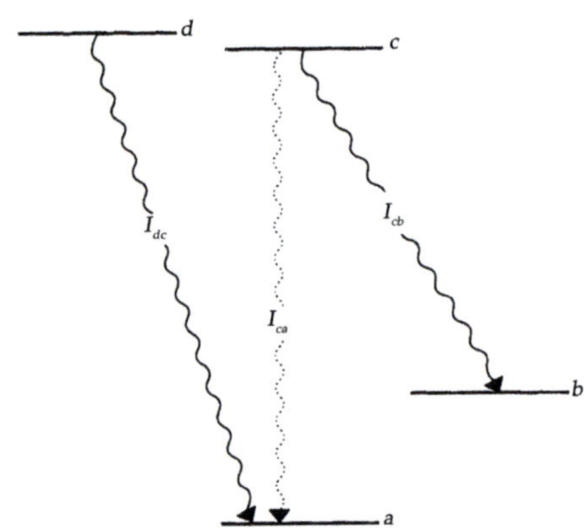

Figure 1.8. *Partial Grotrian diagram of a typical isolated two-state system used to observe the Stark effect. Levels c and d are separated by a small energy gap at zero electric field, and are connected by an electric dipole-allowed transition.*

Source: J.E. Lawl, Creative Commons License.

These types are determined by the electric field strength along with the states involved (Meixner et al., 1986) (Figure 1.8).

1.3.4.1. Linear Stark Effect

The linear Stark effect occurs when the energy shift between the spectrum is proportional to the magnitude of the electrical field that is used.

This phenomenon is typically found in hydrogen-like ions or atoms that possess non-degenerate levels of energy. The change in energy (ΔE) may be mathematically represented as:

$$\Delta E = -\mu E$$

The symbol (μ). represents the dipole electrical moment within an atom or molecule, while (E) represents the electric field intensity (Calvet et al., 2007).

1.3.4.2. Quadratic Stark Effect

The Stark effect has a quadratic relationship among molecules or atoms that possess defective energy levels. The energy change in this instance is directly proportional to the square of the magnitude of the electric field. This quadratic Stark effect may be mathematically represented as:

$$\Delta E = -\frac{1}{2}\alpha E^2$$

where; (α) represents the polarizability of an atom or molecule (Pokutnyi et al., 2004).

1.3.4.3. Perturbation Theory and Stark Effect

An analysis of the Stark effect may be conducted via the perturbation concept. When an atom is subjected to an outside electrical field $(\vec{E})$, its perturbed Hamiltonian (H') can be determined:

$$H' = -\vec{d} \cdot \vec{E}$$

where; $(\vec{d})$ represents the electrical dipole operator. By employing the initial perturbation concept, one can determine the energy shift (ΔE_n) for the (n) –th levels of energy.

$$\Delta E_n^{(1)} = \langle \psi_n | H' | \psi_n \rangle = -\langle \psi_n | \vec{d} \cdot \vec{E} | \psi_n \rangle$$

To account for higher-order effects, it is necessary to use the second-order perturbations theory, which results in a quadratic relationship with the electric field strength (Alvarez & Sundaram, 2004).

1.3.5. Zeeman Effect and Magnetic Field Influence on Spectra

The Zeeman effect refers to the division or displacement of spectral lines when a magnetic field from outside is present. This phenomenon occurs because of the magnetic field interacting with a magnetic dipole moment that is linked to the orbit as well as spin angular momenta for electrons. The Zeeman effect plays a pivotal role in comprehending the magnetic characteristics of both molecules and atoms which is extensively employed in methodologies such as radiation spectroscopy (Morino & Toyama, 1961).

Did you know?

Did you know that the Zeeman effect, observed as the splitting of spectral lines in the presence of a magnetic field, provides insights into the magnetic properties of atoms and is essential for magnetic resonance imaging (MRI) technology?

1.3.5.1. Normal and Anomalous Zeeman Effect

The Zeeman effect may be categorized into two types: the regular Zeeman effect versus the anomalous Zeeman impact. The usual Zeeman effect manifests among atoms or ions possessing a solitary unpaired electron, such as a molecule of hydrogen (Mathys & Stenflo, 1987). This leads to the division of the spectral lines across three distinct components, forming a triplet. The change in energy (ΔE) with the average Zeeman effect can be expressed as:

$$\Delta E = m_l \mu_B B$$

where; (m_l) represents the magnetism quantum number; (μ_B) represents the Bohr magneton; and (B) represents the magnetic field intensity.

The anomalous Zeeman effect manifests in atoms or ions that possess many unpaired electrons and entail intricate splitting patterns resulting from the interaction between electron spin and orbital motion. The energy shift associated with the anomalous Zeeman impact is:

$$\Delta E = g_J \mu_B m_J B$$

where; (g_J) represents the Landé g-factor; (m_J) denotes the overall angular momentum quantum numbers; and (μ_B) along with (B) are defined as mentioned earlier (Werner et al., 1992).

1.3.5.2. Quantum Mechanical Treatment of the Zeeman Effect

The Hamiltonian over an atom in the presence of a magnetic field that is external $(\vec{B})$ can be expressed as:

$$\pm = \ _0 + \ '$$

where; (H_0) represents the unaffected Hamiltonian; and (H') corresponds to the perturbation Hamiltonian caused by a magnetic field. The change in the Hamiltonian (H') may be expressed as:

$$H' = -\vec{\mu} \cdot \vec{B}$$

where; $(\vec{\mu})$ represents the magnetic moment of the dipole. The magnetic dipole moment $(\vec{\mu})$ can be determined for an electron by considering its orbital angular $(\vec{L})$. momentum and spin angular momentum $(\vec{S})$:

$$\vec{\mu} = -\mu_B (g_L \vec{L} + g_S \vec{S})$$

in which (g_L) along with (g_S) represent the g-factors associated with orbital along with spin angular momenta, accordingly (Heisenberg & Jordan, 1926).

1.3.5.3. Energy Shifts and Splitting

Calculating the energy shifts caused by the effect of Zeeman may be performed with the first-order perturbation concept. In the case of the usual Zeeman effect, when only considering the orbital contributions, the change in energy (ΔE) can be determined (Green et al., 2014).

$$\Delta E = \mu_B B m_l$$

When examining the anomalous Zeeman impact, it is important to consider each of the orbitals along with spin contributions to determine the energy shift (ΔE)

$$\Delta E = \mu_B B g_J m_J$$

This Landé g-factor (g_J) can be determined by the following equation:

$$g_J = 1 + \frac{J(J+1) + S(S+1) - L(L+1)}{2J(J+1)}$$

Pressure broadening, Stark effect, or Zeeman impact are basic phenomena that occur when matter interacts with external fields (Darwin, 1926).

1.4. ADVANCED QUANTUM CONCEPTS

1.4.1. Perturbation Theory in Atomic Systems

The perturbation theory is a crucial technique in quantum physics, especially for comprehending systems that cannot be precisely solved. It offers a technique for estimating the eigenvalues or eigenstates for a Hamiltonian which is slightly different from a problem that can be solved exactly (Graff et al., 1985).

When considering a Hamiltonian (H) that consists of a solvable component (H_0) along with a minor perturbation, (H_0):

$$H = H_0 + \lambda H'$$

when (λ) is the smallest parameter. One is familiar with the answers for (H_0), the unaffected system.

$$H_0 \left| \psi_n^{(0)} \right\rangle = E_n^{(0)} \left| \psi_n^{(0)} \right\rangle$$

To be able to determine the first-order adjustments for the energy (E_n), one utilizes:

$$E_n^{(1)} = \left\langle \psi_n^{(0)} | H' | \psi_n^{(0)} \right\rangle$$

This wavefunction $(|\psi_n\rangle)$ can be corrected to first order:

$$\left| \psi_n^{(1)} \right\rangle = \sum_{m \neq n} \frac{\left\langle \psi_m^{(0)} | H' | \psi_n^{(0)} \right\rangle}{E_n^{(0)} - E_m^{(0)}} \left| \psi_m^{(0)} \right\rangle$$

This results in the eigenstate approximation of the first order:

$$|\psi_n\rangle \approx \left| \psi_n^{(0)} \right\rangle + \lambda \left| \psi_n^{(1)} \right\rangle$$

Additional corrections may be derived similarly, but for most practical applications, the first-order perturbation theory is a satisfactory estimation. One application for perturbation theories in atomic processes is its use in studying the fine framework splitting within hydrogen-like atoms. Relativistic modifications and spin-orbit linking result in a fine structure, that can be considered as perturbations towards the non-relativistic Hamiltonian (Sadlej, 1981).

A Hamiltonian equation for a fine structure may be written as (H_{FS}).

$$\pm_{FS} = \quad _{rel} + \quad _{so}$$

The term (H_{rel}) takes into consideration relativistic modifications, while (H_{so}) represents the spin-orbit linking term:

The correction (H_{rel}) was relativistic:

$$H_{rel} = -\frac{p^4}{8m^3c^2}$$

In which (p) represents the momentum of an operator, which is (m) denotes the electron's weight, and (c) signifies the velocity of light:

The spin-orbit linking (H_{so}) can be expressed as:

$$H_{so} = \frac{1}{2m^2c^2}\frac{1}{r}\frac{dV}{dr}\vec{L}\cdot\vec{S}$$

in which $V(r)$ indicates the Coulomb's potential, $(\vec{L})$ corresponds with the orbital momentum, along $(\vec{S})$ represents its spin angular momentum.

When studying a hydrogen atom, one can analyze the fine structure via perturbation theory. This involves treating the relativistic along with spin-orbit modifications as minor perturbations of the non-relativistic Hamiltonian. The overall Hamiltonian incorporating these influences is:

$$H \aleph H_0 \quad H_{rel} \quad H_{so}$$

Here, one has the non-relativistic Hamiltonian denoted as (H_0), a relativistic correction denoted as (H_{rel}), along with a spin-orbit linking term denoted as (H_{so}).

The levels of energy might be stated with a relativistic correction:

$$E_{rel} = -\frac{E_n}{2}\left(\frac{Z\alpha}{n}\right)^2\left[\frac{1}{j+\frac{1}{2}} - \frac{3}{4n}\right]$$

In the given equation, the variables (E_n), $(Z)\,(\alpha)\,(n)$, and (j) represent different quantities related to the energy, fine-structure constant, atomic number, and angular momentum quantum numbers.

The overall energy shift due to fine structure is determined by combining the effects of relativity and spin-orbit interactions:

$\Delta E_{FS} = E_{rel} + E_{so}$ The very close lines within the hydrogen spectrum reveal the intriguing phenomenon of intricate structure splitting (Lindgren, 1978).

1.4.2. Variational Methods for Atomic Energies

The variational method is another powerful technique used to approximate the ground state energy of quantum systems. It is based on the variational principle, which states that for any trial wavefunction (ψ_{trial}), the expectation value of the Hamiltonian is an upper bound to the true ground state energy (E_0).

The variational procedure is a highly effective technique employed to estimate the energy of the lowest state in quantum systems. The idea on which it is founded is called the variational principle, which asserts that the expected value within the Hamiltonian for every trial wavefunction (ø_{trial}), represents a limit for the actual ground state energy (E_0):

$$E0 \leq \langle \psi_{trial} \rangle \mid H \mid \psi_{trial} \rangle \rangle$$

The objective is to select a trial wavefunction (ψ_{trial}) that relies on a few factors and subsequently reduces the expectation value about such parameters (Massey & Moiseiwitsch, 1951).

Let's use the ground condition energy found in a hydrogen atom as an example. An elementary trial wavefunction could be:

$$\psi_{trial}(r) = e^{-\alpha r}$$

in which (α) represents a variational parameter. The average value for the Hamilton (H) becomes

$$\langle H \rangle = \frac{\int \psi^*_{trial}(r) H \, \psi_{trial}(r) \, d^3r}{\int \psi^*_{trial}(r) \psi_{trial}(r) \, d^3r}$$

The Hamiltonian equation for the atom of hydrogen is given by:

$$H = -\frac{\hbar^2}{2m}\nabla^2 - \frac{e^2}{4\pi\epsilon_0 r}$$

By evaluating integrals and then minimizing $\langle H \rangle$ about (α)one may obtain a rough estimate of the base state energy.

- **Variational Methods Applied to the Helium Atom:** The helium atom, consisting of two electrons, is not solvable precisely because of its electron-electron repulsive factor. The variational technique offers a valuable estimation of the energy of the lowest possible state (Yu & Muckerman, 2002). An often-used trial wavefunction over helium involves the multiplication of hydrogen-like orbits with a powerful nuclear charge $(Z_{\pm})$:

$$\psi_{trial} = \psi_{1s}(r_1, Z_{eff})\psi_{1s}(r_2, Z_{eff})$$

The function $(\psi_{1s}(r, Z_{eff}))$represents the hydrogenic 1s orbital, which considers the real nuclear charge (Z_{eff}). The Hamiltonian's expected value reduces about $(Z_{\pm})$:

$$H = -\frac{\hbar^2}{2m}(\nabla_1^2 + \nabla_2^2) - \frac{Ze^2}{4\pi\epsilon_0 r_1} - \frac{Ze^2}{4\pi\epsilon_0 r_2} + \frac{e^2}{4\pi\epsilon_0 |\vec{r_1} - \vec{r_2}|}$$

By calculating integrals and reducing $\langle H \rangle$ concerning (Z_{eff}), one can obtain a rough figure of the ground condition energy of helium (Bowman et al., 2008).

1.4.3. Spin-Orbit Coupling and Fine Structure

Spin-orbit coupling represents a fundamental quantum mechanical phenomenon that occurs because of a relationship with the spin of an electron and how it moves about the nucleus. The interaction between these two components results in the phenomenon known as fine structure splitting, which causes the energy levels within atoms to separate (Subramanian, 1996).

The spin-orbit interaction arises because of the relativisticadjustments made to the Schrödinger equation. This correction considers the magnetic field that the electron experiences in its rest frame because of the orbital motion. The relationship Hamiltonian (H_{so}) can be obtained by:

The potential energy $\xi(r)$e and the dot product for the orbital along with spin angular momenta, $\vec{L}\cdot\vec{S}$, are important concepts in the study of matter in physics. $\xi(r)$ may be expressed over hydrogen-like atoms:

Tip

Spin-orbit coupling, which results from the interaction between an electron's spin and its orbital motion around a nucleus, leads to the fine structure splitting in atomic spectra. Understanding this phenomenon is crucial for high-precision spectroscopy and quantum information science.

$$\xi(r) = \frac{1}{2m^2c^2}\frac{1}{r}\frac{dV}{dr}$$

where; $V(r) = -\frac{Ze^2}{4\pi\epsilon_0 r}$ the Coulomb potential is represented as (Ze). Plugging in such a possibility into $\xi(r)$ yields:

$$\xi(r) = \frac{Ze^2}{8\pi\epsilon_0 m^2c^2r^3}$$

The overall angular momentum $(\vec{J})$ is the combined vector that includes the orbital angular momentum $(\vec{L})$ and the spin angular momentum $(\vec{S})$:

$$\vec{J} = \vec{L} + \vec{S}$$

These eigenstates for the overall angular momentum (J) can be identified from quantum numbers such as (j),(l), or (s) Calculating the energy shift caused by spin-orbit coupling involves determining the expected value of a certain quantity $(\vec{L}\cdot\vec{S})$:

$$\langle\vec{L}\cdot\vec{S}\rangle = \frac{1}{2}[j(j+1) - l(l+1) - s(s+1)]\hbar^2$$

Thus, the energy correction resulting from the interaction between spin and orbital motion is:

$$E_{so} = \langle H_{so}\rangle = \frac{\hbar^2}{2m^2c^2}\frac{1}{r}\frac{dV}{dr}\frac{1}{2}[j(j+1) - l(l+1) - s(s+1)]$$

When considering hydrogen-like atoms:

$$E_{so} = \frac{Z^4 e^4}{8\pi^2 \epsilon_0^2 \hbar^3 c^2}\left(\frac{1}{n^3}\right)\left[j(j+1) - l(l+1) - \frac{3}{4}\right]$$

The energy correction results in the division of the energies into smaller components for fine structure, that can be identified by the atomic spectra.

- **Exploring the Multi-Electron Atoms:** Interactions among multiple electrons with multi-electron atoms make spin-orbit linking more intricate. The complete Hamiltonian, which considers the effects of spin-orbit coupling, seems:

$$H = H_0 + \sum_i H_{so}^{(i)}$$

where; (H_0) indicates an unperturbed Hamiltonian along with $\left(H_{so}^{(i)}\right)$ denotes the spin-orbit coupling terms that define the ith electron. The spin-orbit linking terms for every single electron can be calculated.

$$H_{so}^{(i)} = \xi(r_i)\vec{L_i} \cdot \vec{S_i}$$

The overall spin-orbit coupling energy arises from the combined effects of each of the electrons:

$$E_{so} = \sum_i \left\langle H_{so}^{(i)} \right\rangle$$

Aside from spin-orbit coupling, it is crucial to consider various interactions like electron-electron repelling and exchanging interactions. The interactions between particles result in intricate splitting patterns within the energy levels, referred to as both fine and hyperfine frameworks (Machala et al., 2007) (Figure 1.9).

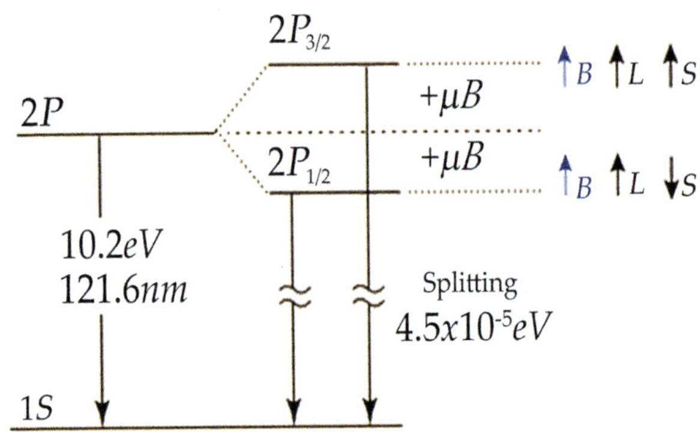

Figure 1.9. *Hydrogen fine structure.*

Source: Hyper Physics, Creative Commons License.

Recognizing the intricate actions of atomic systems requires a grasp of advanced quantum ideas like perturbation theory, variational methods, spin-orbit coupling, and spin-orbit coupling. These techniques offer estimations for the energy levels along with atom wavefunctions, enabling us to analyze and forecast spectral lines as well as other phenomena (Magonov, 1993).

1.5. EXPERIMENTAL TECHNIQUES

1.5.1. Spectroscopy Methods for Atomic Analysis

Spectroscopy is a highly effective method for studying the characteristics of atoms and molecules through the analysis of their interaction using electromagnetic radiation. There are various spectroscopic techniques available, each offering unique insights into the sample. There are several commonly employed procedures in spectroscopy, including emission, absorption, and scattering techniques (Subramanian, 1996).

1.5.1.1. Absorption Spectroscopy

Measurement of light absorption through a sample is done through absorption spectroscopy, which examines the absorption as an indicator of wavelength. Atoms or molecules can absorb wavelengths that light emits that correspond with the difference in energy among two quantum states. The Beer-Lambert law explains how light is absorbed when it travels through a substance:

$$A = \log\left(\frac{I_0}{I}\right) = \epsilon c l$$

The absorbance is represented by (A), the incoming light intensity is denoted by (I_0), the emitted light intensity by (I), the molar absorptivity by (ϵ), the concentration for absorbing material through (c), and the pathway that carries the light via the specimen is denoted by (l).

Atomic absorption spectroscopy (AAS) involves measuring the absorbed energy of visible light through atoms within the gas phase. This method is commonly employed to quantitatively analyze metallic elements (Winefordner et al., 2004).

1.5.1.2. Emission Spectroscopy

Emission spectroscopy is a field that focuses on studying the light emitted through atoms or molecules when they are energized at higher energy levels. While the excited particles return to their initial levels of energy, they release light at wavelengths that match the energy variation among the states.

The relationship between the intensity of the light that is released and the concentration of the emitting species could be determined using a specific equation:

$$I = kc$$

The equation represents the relationship between the intensity of emitted light (I), a value of proportionality (k), and the concentration within the emitting species (c).

Flame emission spectrometers including inductively coupled plasma emission spectroscopy (ICP) have become common methods for analyzing the chemical composition of materials (Aragón & Aguilera, 2008).

1.5.1.3. Scattering Spectroscopy

Scattering spectroscopy, which includes Raman spectroscopy, quantifies the dispersion of light caused by molecules. Raman spectroscopy involves the interaction of incident light by molecular vibrations, resulting in the emission of scattered light that has a frequency shift matching the energy values of the molecules' vibrations. This Raman shift $\Delta\nu$ is determined by:

$$\Delta\nu = \nu_i - \nu_s$$

where; (ν_i) represents the frequency of the incident light; and (ν_s) represents the wavelength of the dispersed light.

The magnitude of a Raman signal is directly proportional to the number of the species that cause the scattering.

$$I_R = Kc$$

where; (I_R) indicates the Raman intensity; (K) serves as a proportionality constant; and (c) denotes the concentration for the dispersing species (Backman et al., 2001) (Figure 1.10).

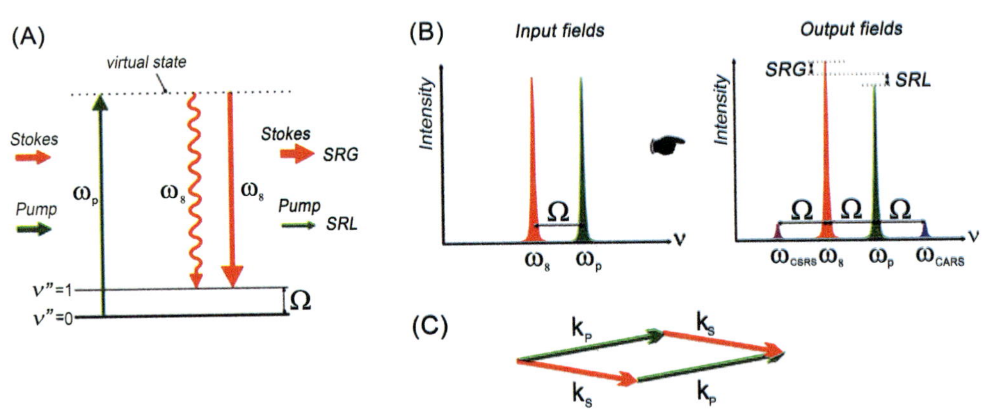

Figure 1.10. *(A) Energy level presentation for stimulated Raman scattering (SRS) process, (B) spectral positions of input fields and the generated output fields, and (C) a representation showing phase matching automatically satisfied in SRS. CARS, coherent anti-Stokes Raman scattering; CSRS, coherent Stokes Raman scattering; SRG, stimulated Raman gain; SRL, stimulated Raman loss.*

Source: Vikas Kumar, Creative Commons License.

1.5.2. Quantum Interference and Coherence Effects

Understanding the principles of quantum interference along with coherence is crucial in the field of quantum mechanics, including far-reaching effects on various experimental methods and technological advancements.

1.5.2.1. Quantum Interference

While quantum particles overlap, quantum interference takes place. This phenomenon can result in either constructive or destructive interference patterns. You can witness this phenomenon in a range of experiments, including the double-slit experiment, where particles including electrons display wave-like interference patterns (Kornberg et al., 2012).

The calculation of the probability amplitude over a particle that crosses two slits (A) along with (B) can be determined by the following equation:

$$\Psi = \Psi_A + \Psi_B$$

Next, one has the probability density (P):

$$P = |\Psi|^2 = |\Psi_A + \Psi_B|^2$$

Upon further examination of this expression, one can derive:

$$P = |\Psi_A|^2 + |\Psi_B|^2 + \Psi_A\Psi_B^* + \Psi_A^*\Psi_B$$

The terms involving the wave functions $(\Psi_A \Psi_B * + \Psi_A * \Psi_B)$ can either lead to constructive or destructive interference, according to the phase variance between (Ψ_A) and (Ψ_B).

1.5.2.2. Coherence Effects

Coherence relates to the consistent phase connection between waves, including light waves along with matter waves. Observing the impacts of interference requires coherence, which is determined by the length and time of coherence.

Light sources that are coherent, like lasers, possess a precise phase relationship along can sustain coherence over significant distances. On the other hand, sources that lack coherence, such as thermal light, exhibit unpredictable phase relationships (Simon et al., 2015).

The length (L_c) that describes the distance over which coherence is maintained is provided by:

$$L_c = \frac{\lambda^2}{\Delta\lambda}$$

where; (λ) represents the wavelength that emits light; and $(\Delta\lambda)$ denotes the spectral width of the light source.

The relationship between the time of coherence (τ_c) and the coherence width is given by:

$$\tau_c = \frac{L_c}{c}$$

where; (c) represents the velocity of light (McNamara, 2001).

1.5.3. Application and Advanced Techniques

Atomic Force Microscopy (AFM) and Scanning Tunneling Microscopy (STM) are highly effective methods for visualizing surfaces at the atomic level. Both methods offer exceptional image quality and can be utilized to examine the surface characteristics and attributes of materials (Li et al., 2012).

1.5.3.1. Atomic Force Microscopy (AFM)

AFM can measure the forces among a sharp tip along with the outer layer of a sample, allowing for the generation of images with a high resolution.

The AFM tip is connected to a cantilever, that bends when it interacts using the surface. The measurement of deflection involves the utilization of a laser beam that is reflected off the rear part of the cantilever and directed toward a photodetector (Goldsbury et al., 2009).

The forces detected in AFM encompass a range of interactions, such as electrostatic forces, van der Waals forces, and chemical bonding forces. The interaction between the tip along the surface may be explained using Hooke's law:

$$F = -k\,\Delta z$$

The variables in the equation are (F) with force, (k) for the spring constant that governs the cantilever, while and (Δz) for the flexion angle from the cantilever.

AFM functions in several modes, including contact mode, non-contact mode, and tapping mode.

During contact mode, the tip maintains constant contact via the surface, yet during the non-contact mode, its tip hovers above the surface and does not make physical contact. Tapping mode is a technique where the tip makes intermittent contact with the surface (Fang et al., 2000).

1.5.3.2. Scanning Tunneling Microscopy (STM)

STM is based on the phenomenon of quantum tunneling, where electrons can pass through barriers and move among the sharp metallic tip along with the surface of the sample. When a voltage is placed along the tip along with the sample, electrons can pass using the vacuum gap, resulting in the generation of a tunneling current. The relationship between the electrical current and the space between the tip along the surface is rapid:

$$I \propto e^{-\kappa d}$$

where; (I) represents the tunneling current; (δ) represents a constant that is associated with the barrier's height; and (d) represents the distance between the tip and the surface.

The tunneling current is highly dependent on the distance between the tip and the sample, which enables the scanning tunneling microscope (STM) to achieve atomic-level resolution. An image of the surface's topography is generated by recording the current while raster-scanning the STM tip through the surface. AFM and STM have undergone significant advancements to enable a range of advanced methods, such as atomic-scale spectroscopy and manipulation (Binnig & Rohrer, 1987).

1.5.3.3. AFM Spectroscopy

AFM is a powerful tool for analyzing the characteristics of different materials by studying force-distance spectroscopy.

Take, for instance, the analysis of the force-distance curve to ascertain the Young's modulus (E) from a material.

$\overline{\text{ä}}$

where; (F) represents the force exerted on the AFM tip along with (ä) denotes the depth of indentation (Dazzi et al., 2012).

1.5.3.4. STM Spectroscopy

STM is a powerful tool that allows for the analysis of electronic structures through scanning tunneling spectroscopy (STS). Through the measurement of the different conductance $\left(\frac{dI}{dV}\right)$ concerning the used bias voltage (V), it is possible to determine the local density of states (LDOS) for the sample.

This tunneling current (I) can be calculated by the local density of states (LDOS) $(\tilde{n}(E))$ for the sample.

$$I(V) \propto \int_{-\infty}^{eV} \rho(E)T(E,eV)\, dE$$

The variable $T(E,eV)$ represents the likelihood of electrons successfully tunneling past the barrier.

Scanning Tunneling Spectroscopy (STS) can provide detailed information regarding the electronic states, energy gaps, and many more electronic features of the sample with atomic-level precision.

1.6. APPLICATIONS IN MODERN PHYSICS

The uses in current physics are explained in the following sections.

1.6.1. Atomic Clocks and Precision Measurement

Atomic clocks are renowned for their unparalleled accuracy in keeping time, dependent on the meticulous measurement of electrical changes within atoms. These instruments rely on the regular vibrations of atoms to be a reference for measuring time. One of the frequently used atomic clocks includes the cesium nuclear clock, which depends on a change from two hyperfine states that constitute the fundamental state of cesium-133 atoms (Bauch, 2003).

1.6.1.1. Principles of Atomic Clocks

An atomic clock operates based on microwaved transitions that occur within the hyperfine energies of an atom. The phase change for cesium-133 happens at a rate of around 9,192,631,770 Hz. This frequency is the defining factor for the second part of (SI).

This hyperfine splitting (ΔE) within the ground condition of cesium-133 can be determined using the following equation:

$$\Delta E = h\nu$$

In this equation, the energy variance among both hyperfine levels is denoted as (ΔE) Planck's constant is represented by (h) along the transition frequency is denoted as (ν)

Within a cesium nuclear clock, a stream of cesium atomic particles is directed via a microwave cavity, so they encounter microwave radiation with a frequency near 9,192,631,770 Hz.

The atoms start in a particular state, and when the frequency of the microwave fits the change in frequency, that one will take in energy or switch to a different state. The measurement of the atoms involved in this transition is taken, and the frequency for the microwave energy is fine-tuned to optimize this measurement, guaranteeing the clock's precision (Gabriel & Haugan, 1990).

1.6.1.2. Accuracy and Stability

Atomic clocks rely on accurate measurements of transition frequencies and their capacity to minimize external disturbances like magnetic fields along with temperature changes to maintain their accuracy. The clock's stability is determined through the Allan deviation $(\sigma_y(\tau))$, which quantifies a fraction frequency instability during a specific time interval (τ)

$$\sigma_y(\tau) = \sqrt{\frac{1}{2}\left\langle\left(\frac{\nu_i - \nu_{i+1}}{\nu_0}\right)^2\right\rangle}$$

Given the observed frequencies on subsequent intervals, denoted as (ν_i) or (ν_{i+1}) along with the nominal frequency, (ν_0) one can analyze the relationship between these values.

Innovative atomic clocks, like optical lattice clocks that utilize strontium along with ytterbium atoms, have achieved remarkable levels of precision and consistency, boasting uncertainties on the scale of (10^{-18}) or even smaller. These clocks have important uses in various fields like global positioning systems (GPS), telecommunications, and research within the fundamental properties of matter (Berth & Jeschke, 1989).

1.6.2. Quantum Computing with Atomic Systems

Quantum computing utilizes the concepts of quantum physics to carry out calculations that are beyond the capabilities of traditional computers. Trapped ions along with neutral atoms have emerged as highly promising options for constructing quantum computers. These atomic systems possess precise quantum states and exhibit remarkable coherence that can be sustained for extended durations (Bresler & Dagan, 1983).

1.6.2.1. Qubits and Quantum Gates

Quantum computers operate using a fundamental building block called the qubit. These qubits have the unique ability to exist within a superposition from states, represented as $(|0\rangle)$ and $(|1\rangle)$ Qubits in atomic systems are commonly depicted through the internal electronic configurations of atoms or ions (Ivanov et al., 2015).

A wavefunction may be used to describe the record of a qubit $(|\psi\rangle)$:

$$|\psi\rangle = \alpha|0\rangle + \beta|1\rangle$$

where; (α) along with (β) are coefficients that have complex values and satisfy $(|\alpha|^2 + |\beta|^2 = .1$

Quantum gates are utilized for controlling qubits, enabling the implementation of unitary changes in their current states. Some of the commonly used gates for single qubits are the Pauli-X gate, the Hadamard gate, along the phase gate. Multi-qubit gates, including the controlled-NOT (CNOT) gate, allow for the creation of entanglement among qubits (Plantenberg et al., 2007).

The effect that a quantum gate (U) upon a qubit state $(|ø)$ can be described as:

$$U\,|\psi\rangle = U\,(\alpha|0\rangle + \beta|1\rangle)$$

As an illustration, a Hadamard gate H) alters the basis states in the following manner:

$$H\,|0\rangle = \frac{1}{\sqrt{2}}(|0\rangle + |1\rangle)$$

$$H\,|1\rangle = \frac{1}{\sqrt{2}}(|0\rangle - |1\rangle)$$

1.6.2.2. Trapped Ions and Neutral Atoms

Trapped ions along with neutral atoms possess unique benefits for quantum computing. For trapped ion structures, ions are contained using electromagnetic radiation within a vacuum chamber. The qubit levels are denoted by hyperfine and optical changes of the ions. Resonate interaction is employed to execute quantum gates using laser pulses (Cirac & Zoller, 2004).

The Hamiltonian describing the interaction between a trap ion qubit and a laser field can be expressed as:

$$H = \frac{\hbar\omega_0}{2}\sigma_z + \hbar\Omega(t)\left(\sigma_+ e^{-i\omega t} + \sigma_- e^{i\omega t}\right)$$

In this context, (ω_0) represents the transition frequency of qubit, $\Omega(t)$ corresponds to the Rabi frequency in the laser field, along with $(\sigma_z),(\sigma_+)$ or (σ_-) define the Pauli matrices.

Neutral atoms may be confined by optical lattices or optical tweezers, generating periodic potential wells using laser light. The states that define qubits are encoded within the ground and high states for the atoms. Transitions among states can be induced via laser pulses and microwave fields when creating quantum gates (Dotsenko et al., 2005).

The Hamiltonian equation over an atom with no charge qubit within an optical lattice can be expressed as:

$$[H = \sum_i \left(\frac{\hbar^2}{2m}\nabla_i^2 + V(r_i)\right) + \sum_{i<j} U(r_i - r_j)]$$

In the equation, the mass of the atom is denoted by (m), the optical lattice potential is represented by $V(r_i)$, and the reaction potential within atoms is given by $U(r_i - r_j)$.

1.6.2.3. Error Correction and Scalability

Understanding the core principles of quantum error mitigation is crucial in the development of functional quantum computers. This is because qubits, the building blocks of quantum computing, are highly vulnerable to decoherence and several types of errors. One way to safeguard quantum information is by utilizing error correction codes, like the surface code. This method involves encoding logical qubits through numerous physical qubits and employing redundancy to identify and rectify errors.

The surface-based code utilizes a two-dimensional array for qubits, where stabilizer measures are conducted to identify and address errors. The logic of each qubit is determined by the current state for the whole array, and any errors that occur are rectified by implementing the necessary unitary operations. Quantum computing faces a significant obstacle of scalability. Atomic systems present a favorable approach to achieving scalability because of their inherent consistency and the capability to manipulate many qubits using laser or microwave fields. Progress into ions trap technology, optical lattices, and integrated photonics is facilitating the development of quantum computers on a massive scale, capable of accommodating thousands or millions of qubits (Terhal et al., 2020).

1.6.3. Atomic Models in Astrophysics and Cosmology

Atomic models are essential in astrophysics including cosmology as they offer a valuable understanding of the chemical reactions and processes that take place in stars, galaxies, especially the interstellar medium (Arkani et al., 1999).

1.6.3.1. Stellar Spectroscopy and Atomic Absorption Lines

Stellar spectroscopy entails the examination of the light released or taken in by stars to ascertain their chemical makeup, temperature, density, and further characteristics. The spectra near stars exhibit absorption lines that correspond to the transformations of both ions and atoms within the atmosphere around a star.

Various parameters, including the temperature as well as pressure within the stellar environment, influence the intensity along with the shape of lines that absorb light. The Boltzmann distribution characterizes the arrangement of atomic energies inside a system at a specific temperature:

$$\frac{N_j}{N_i} = \frac{g_j}{g_i} e^{-(E_j - E_i)/k_B T}$$

The variables (N_j) and (N_i) represent the population sizes (j) and (i), respectively. (g_j) and (g_i) represent the degeneracies of these levels. (E_j) represent their energies. (k_B) represents Boltzmann's constant while (T) represents the temperature.

The Saha equation elucidates the state of ionization equilibrium within a star environment:

$$\frac{N_{i+1} N_e}{N_i} = \frac{2\Lambda^3}{g_i} \left(\frac{2\pi m_e k_B T}{h^2}\right)^{3/2} e^{-\chi_i/k_B T}$$

Let (N_{i+1}) and (N_i) represent the populations for ionized and neutral atoms, respectively. (N_e) denotes the electron density, (Λ) stands for the partition function (g_i) represents the statistical weight, (m_e) signifies the electron mass (χ_i) denotes the ionization energy, and (h) represents Planck's constant (Fowler, 1925).

1.6.3.2. Atomic Processes in the Interstellar Medium

This interstellar medium (ISM) consists of gas and dust located between stars, where atomic or molecular mechanisms have a substantial impact on its dynamics along with evolution. Atomic models provide a framework for understanding and explaining various phenomena (Kroto, 1988), including emission lines, absorption lines, including ionization states for atoms in the interstellar medium (ISM).

The radiative flow equation shows how radiation moves through the ISM:

$$\frac{dI_\nu}{ds} = j_\nu - \alpha_\nu I_\nu$$

In the given equation, (I_ν) represents the specific amount of radiation during a particular frequency (ν), (s) denotes the path's length, (j_ν) represents the emission factor, and (α_ν) stands for the absorbed coefficient. The ionization of hydrogen in the interstellar medium is explained through the Strömgren sphere, which is a region of hydrogen that becomes ionized due to the presence of a nearby hot star. The Strömgren sphere's radius, (R_S) can be determined using the following equation:

$$R_S = \left(\frac{3S_*}{4\pi\alpha_B n_e^2}\right)^{1/3}$$

where; (S_*) represents the ion photon flux emitted by the star, (α_B) denotes the amount of recombination coefficients, and (n_e) represents the electron density (Roueff & Lique, 2013).

1.6.3.3. Atomic Models in Cosmology

Atomic models play a crucial role in unraveling the mysteries of cosmology, shedding light on the origins of the cosmos, the cosmic microwave background (CMB), along with the complex method of element formation known as nucleosynthesis.

During the recombination epoch, there was a significant event where electrons and protons came together to create neutral hydrogen. This allowed photons to break free from matter and move around without any hindrance, leading to the formation of the cosmic microwave background (CMB). The ionization course of the Universe is explained via the Saha equation alongside the equations for recombination rates (Bunkov, 2008).

The baryon-to-photon proportion \ (η\) is a crucial parameter in the field of cosmology, since it directly affects the number of elements that are formed during the process of Big Bang nucleosynthesis. The initial quantities of basic elements, including hydrogen, helium, and lithium, offer limitations on cosmological theories and the circumstances of the early cosmos.

The Planck spectrum characterizes the blackbody radiation emitted by the cosmic microwave background (CMB):

$$B_\nu(T) = \frac{2h\nu^3}{c^2}\frac{1}{e^{h\nu/k_B T} - 1}$$

The equation represents the relationship between the spectral brightness frequency, (ν) (B_ν), and temperature (T) speed of light (c) and Planck's constant (h).

Atomic models are essential for comprehending the absorption and emission lines detected in the spectra of faraway galaxies and quasars. They offer valuable information about the chemical composition, temperature, and ionization states of the intergalactic medium (Uzan, 2011).

Modern physics finds critical applications in nuclear clocks, quantum computing using atomic systems, along with atomic frameworks for astrophysics and cosmology. These fields utilize the accurate and fundamental comprehension of atomic systems to progress technology, investigate the universe, and enhance our understanding of the fundamental rules guiding the physical world. Through the utilization of atomic systems, scientists can attain unparalleled precision in measurements, create groundbreaking computational methods, and unveil the enigmas of the universe (McCrea, 1953).

ACTIVITY 1.1.

Explore quantum interference phenomena using optical setups or computational simulations. Investigate how coherence effects influence the behavior of atomic systems, demonstrating applications in quantum computing and precision measurement.

SUMMARY

- This chapter provides a detailed examination of the core concepts, interactions, and applications in atomic and quantum physics. It starts by introducing the basic elements of matter, focusing on the properties of electrons and the composition of nuclei. The interactions between these components are explained using Schrödinger's equation, which is fundamental for understanding microscopic interactions and dynamics in quantum systems.
- The chapter then discusses typical scales that characterize matter, including the interaction energy of elementary point charges and the timescales and frequencies of electronic motions. Key constants, such as electron velocity and the fine-structure constant, are introduced to illustrate the spatial and temporal dimensions within atomic structures. The relationship between electron length and frequency scales offers a comprehensive view of matter, from subatomic particles to complex molecular structures.
- An in-depth exploration of atomic spectra and broadening phenomena follows, covering decay rates and various mechanisms such as Doppler broadening, pressure broadening, collisional effects, the Stark effect, and the Zeeman effect. These discussions demonstrate how external factors and fields influence atomic spectra, affecting their properties under different physical conditions.
- The chapter advances into more complex quantum concepts, including perturbation theory applied to atomic systems, variational methods for calculating atomic energies, and the role of spin-orbit coupling in fine structure. These theoretical frameworks offer deeper insights into the complexities of atomic behavior, energy levels, and transitions.
- Practical applications are highlighted through discussions on experimental techniques like spectroscopic methods for atomic analysis, quantum interference, and coherence effects. These methodologies connect theoretical insights with practical applications, enabling precise measurement, manipulation, and control at the atomic and subatomic levels.
- Finally, the chapter explores significant applications of atomic physics across various fields, including precision timekeeping with atomic clocks, the potential

of quantum computing using atomic systems, and the role of atomic models in enhancing our understanding of astrophysics and cosmology. These applications illustrate the profound impact of atomic physics in advancing technology and expanding our understanding of fundamental physical laws.

REVIEW QUESTIONS

1. What are the basic properties of electrons and how do they contribute to atomic structure?
2. Explain the composition of atomic nuclei and the interactions among their components.
3. Describe Schrödinger's equation and its significance in understanding microscopic interactions.
4. Discuss the different timescales and frequencies associated with electronic motions.
5. How do electric and magnetic fields influence spectral lines?
6. What are perturbation theory and variational methods, and how are they applied in atomic systems?
7. Describe some advanced spectroscopy techniques and their applications in atomic analysis.

MULTIPLE CHOICE QUESTIONS

1. What is the primary function of Schrödinger's equation in quantum mechanics?

a. To describe the motion of macroscopic objects

b. To provide a model for nuclear decay

c. To describe the wavefunction of a quantum system

d. To calculate the temperature of a system

2. Which of the following effects is caused by the influence of an electric field on spectral lines?

a. Doppler effect

b. Stark effect

c. Zeeman effect

d. Photoelectric effect

3. The fine-structure constant is associated with which of the following?

a. Gravitational interactions

b. Strong nuclear force

c. Electromagnetic interactions

d. Weak nuclear force

4. **What experimental technique is commonly used for atomic analysis through the study of light absorption and emission?**
   a. Chromatography
   b. Spectroscopy
   c. Calorimetry
   d. Electrophoresis
5. **In the context of quantum computing, which atomic system is often utilized?**
   a. Atomic clocks
   b. Bose-Einstein condensates
   c. Ion traps
   d. Superconductors
6. **Which broadening mechanism involves the random motion of atoms causing a spread in observed spectral lines?**
   a. Stark broadening
   b. Doppler broadening
   c. Pressure broadening
   d. Natural broadening
7. **The Zeeman effect involves the splitting of spectral lines due to the presence of:**
   a. Electric fields
   b. Magnetic fields
   c. High pressures
   d. Low temperatures
8. **Spin-orbit coupling primarily affects which property of an atom:**
   a. Mass
   b. Charge
   c. Angular momentum
   d. Temperature

Answers to Multiple Questions

1. (c); 2. (b); 3. (c); 4. (b); 5. (c); 6. (b); 7. (b); 8. (c).

REFERENCES

1. Agha, I., Ates, S., Sapienza, L., & Srinivasan, K. (2014). Spectral broadening and shaping of nanosecond pulses: Toward shaping of single photons from quantum emitters. *Optics Letters, 39*(19), 5677–5680. https://doi.org/10.1364/OL.39.005677.

2. Alvarez, G., & Sundaram, B. (2004). Perturbation theory for the Stark effect in a double δ quantum well. *Journal of Physics A: Mathematical and General, 37*(41), 9735. https://doi.org/10.1088/0305-4470/37/41/008.
3. Anazawa, M., Ishikawa, A., Suzuki, T., & Tomoyose, M. (2004). Fractal structure with a typical scale. *Physica A: Statistical Mechanics and Its Applications, 335*(3–4), 616–628. https://doi.org/10.1016/j.physa.2003.12.031.
4. Andelman, D. (2006). Introduction to electrostatics in soft and biological matter. In W. C. K. Poon & D. Andelman (Eds.), *Soft Condensed Matter Physics in Molecular and Cell Biology* (pp. 97–122). CRC Press.
5. Aragón, C., & Aguilera, J. A. (2008). Characterization of laser induced plasmas by optical emission spectroscopy: A review of experiments and methods. *Spectrochimica Acta Part B: Atomic Spectroscopy, 63*(9), 893–916. https://doi.org/10.1016/j.sab.2008.05.010.
6. Arkani-Hamed, N., Dimopoulos, S., & Dvali, G. (1999). Phenomenology, astrophysics, and cosmology of theories with submillimeter dimensions and TeV scale quantum gravity. *Physical Review D, 59*(8), 086004. https://doi.org/10.1103/PhysRevD.59.086004.
7. Backman, V., Gopal, V., Kalashnikov, M., Badizadegan, K., Gurjar, R., Wax, A., & Feld, M. S. (2001). Measuring cellular structure at sub micrometer scale with light scattering spectroscopy. *IEEE Journal of Selected Topics in Quantum Electronics,* 7(6), 887–893. https://doi.org/10.1109/2944.983275.
8. Bauch, A. (2003). Caesium atomic clocks: Function, performance and applications. *Measurement Science and Technology, 14*(8), 1159–1173. https://doi.org/10.1088/0957-0233/14/8/302.
9. Bekenstein, J. D. (2002). Fine-structure constant variability, equivalence principle, and cosmology. *Physical Review D, 66*(12), 123514. https://doi.org/10.1103/PhysRevD.66.123514.
10. Bernard, R., Goutte, H., Gogny, D., & Younes, W. (2011). Microscopic and nonadiabatic Schrödinger equation derived from the generator coordinate method based on zero-and two-quasiparticle states. *Physical Review C, 84*(4), 044308. https://doi.org/10.1103/PhysRevC.84.044308.
11. Berth, P., & Jeschke, P. (1989). Consumption and fields of application of LAS. *Tenside Surfactants Detergents, 26*(2), 75–79.
12. Binnig, G., & Rohrer, H. (1987). Scanning tunneling microscopy—From birth to adolescence. *Reviews of Modern Physics, 59*(3), 615–625. https://doi.org/10.1103/RevModPhys.59.615.
13. Biraben, F., Cagnac, B., & Grynberg, G. (1974). Experimental evidence of two-photon transition without Doppler broadening. *Physical Review Letters, 32*(12), 643–646. https://doi.org/10.1103/PhysRevLett.32.643.
14. Bowman, J. M., Carrington, T., & Meyer, H. D. (2008). Variational quantum approaches for computing vibrational energies of polyatomic molecules. *Molecular Physics, 106*(16–18), 2145–2182. https://doi.org/10.1080/00268970802285720.

15. Bresler, E., & Dagan, G. (1983). Unsaturated flow in spatially variable fields: 2. Application of water flow models to various fields. *Water Resources Research, 19*(2), 421–428. https://doi.org/10.1029/WR019i002p00421.

16. Brus, L. E. (1984). Electron–Electron and electron-hole interactions in small semiconductor crystallites: The size dependence of the lowest excited electronic state. *The Journal of Chemical Physics, 80*(9), 4403–4409. https://doi.org/10.1063/1.447218.

17. Bunkov, Y. M. (2008). 3He: Cosmological and atomic physics experiments. *Philosophical Transactions of the Royal Society A: Mathematical, Physical and Engineering Sciences, 366*(1877), 2821–2832. https://doi.org/10.1098/rsta.2008.0079.

18. Calvet, L. E., Wheeler, R. G., & Reed, M. A. (2007). Observation of the linear Stark effect in a single acceptor in Si. *Physical Review Letters, 98*(9), 096805. https://doi.org/10.1103/PhysRevLett.98.096805.

19. Castleman, A. W., & Bowen, K. (1996). Clusters: Structure, energetics, and dynamics of intermediate states of matter. *The Journal of Physical Chemistry, 100*(31), 12911–12944. https://doi.org/10.1021/jp9601534.

20. Castro Neto, A. H., Guinea, F., Peres, N. M., Novoselov, K. S., & Geim, A. K. (2009). The electronic properties of graphene. *Reviews of Modern Physics, 81*(1), 109–162. https://doi.org/10.1103/RevModPhys.81.109.

21. Chabardes-Garonne, D., Méjean, A., Aude, J. C., Cheval, L., Di Stefano, A., Gaillard, M. C., & Elalouf, J. M. (2003). A panoramic view of gene expression in the human kidney. *Proceedings of the National Academy of Sciences, 100*(23), 13710–13715. https://doi.org/10.1073/pnas.2135133100.

22. Chen, Y. P., Wei, E. I., Jiang, L., Meng, M., Wu, Y. M., & Chew, W. C. (2017). A unified Hamiltonian solution to Maxwell–Schrödinger equations for modeling electromagnetic field–Particle interaction. *Computer Physics Communications, 215*, 63–70. https://doi.org/10.1016/j.cpc.2017.01.003.

23. Cirac, J. I., & Zoller, P. (2004). New frontiers in quantum information with atoms and ions. *Physics Today, 57*(3), 38–44. https://doi.org/10.1063/1.1710367.

24. Claveau, C., Henry, A., Hurtmans, D., & Valentin, A. (2001). Narrowing and broadening parameters of H2O lines perturbed by He, Ne, Ar, Kr and nitrogen in the spectral range 1850–2140 cm−1. *Journal of Quantitative Spectroscopy and Radiative Transfer, 68*(3), 273–298. https://doi.org/10.1016/S0022-4073(00)00125-8.

25. Danos, M., & Geschwind, S. (1953). Broadening of microwave absorption lines due to wall collisions. *Physical Review, 91*(5), 1159–1161. https://doi.org/10.1103/PhysRev.91.1159.

26. Darwin, C. G. (1926). XI. Mechanical models for the Zeeman effect. *The London, Edinburgh, and Dublin Philosophical Magazine and Journal of Science, 1*(1), 161–195. https://doi.org/10.1080/14786442608564103.

27. Dazzi, A., Prater, C. B., Hu, Q., Chase, D. B., Rabolt, J. F., & Marcott, C. (2012). AFM–IR: Combining atomic force microscopy and infrared spectroscopy for nanoscale chemical characterization. *Applied Spectroscopy, 66*(12), 1365–1384. https://doi.org/10.1366/12-06804.

28. Dotsenko, I., Alt, W., Khudaverdyan, M., Kuhr, S., Meschede, D., Miroshnychenko, Y., & Rauschenbeutel, A. (2005). Sub micrometer position control of single trapped neutral atoms. *Physical Review Letters, 95*(3), 033002.

29. Einstein, A. (1936). Physics and reality. *Journal of the Franklin Institute, 221*(3), 349–382.

30. Fang, H. H., Chan, K. Y., & Xu, L. C. (2000). Quantification of bacterial adhesion forces using atomic force microscopy (AFM). *Journal of Microbiological Methods, 40*(1), 89–97.

31. Fowler, R. H. (1925). Notes on the theory of absorption lines in stellar spectra. *Monthly Notices of the Royal Astronomical Society, 85*, 970.

32. Gabriel, M. D., & Haugan, M. P. (1990). Testing the Einstein equivalence principle: Atomic clocks and local Lorentz invariance. *Physical Review D, 41*(10), 2943.

33. Gianfrani, L. (2016). Linking the thermodynamic temperature to an optical frequency: Recent advances in Doppler broadening thermometry. *Philosophical Transactions of the Royal Society A: Mathematical, Physical and Engineering Sciences, 374*(2064), 20150047.

34. Goldsbury, C. S., Scheuring, S., & Kreplak, L. (2009). Introduction to atomic force microscopy (AFM) in biology. *Current Protocols in Protein Science, 58*(1), 17–27.

35. Graffi, S., Grecchi, V., & Silverstone, H. J. (1985). Resonances and convergence of perturbation theory for N-body atomic systems in external AC-electric field. *Annales de l'IHP Physique Théorique, 42*(3), 215–234.

36. Green, J. A., Gray, M. D., Robishaw, T., Caswell, J. L., & McClure-Griffiths, N. M. (2014). A quantum mechanical approach to establishing the magnetic field orientation from a maser Zeeman profile. *Monthly Notices of the Royal Astronomical Society, 440*(4), 2988–2996.

37. Haensel, P., & Zdunik, J. L. (2003). Nuclear composition and heating in accreting neutron-star crusts. *Astronomy & Astrophysics, 404*(2), L33–L36.

38. Halbritter, J., Repphun, G., Vinzelberg, S., Staikov, G., & Lorenz, W. J. (1995). Tunneling mechanisms in electrochemical STM—Distance and voltage tunneling spectroscopy. *Electrochimica Acta, 40*(10), 1385–1394.

39. Heisenberg, W., & Jordan, P. (1926). Application of quantum mechanics to the problem of the anomalous Zeeman effect. *Zeitschrift für Physik, 37*, 263–277.

40. Hindmarsh, W. R., & Farr, J. M. (1973). Collision broadening of spectral lines by neutral atoms. *Progress in Quantum Electronics, 2*, 141–214.

41. Hindmarsh, W. R., Petford, A. D., & Smith, G. (1967). Interpretation of collision broadening and shift in atomic spectra. *Proceedings of the Royal Society of London. Series A. Mathematical and Physical Sciences, 297*(1449), 296–304.

42. Huhtamo, E. (2008). Penetrating the peristrephic: An unwritten chapter in the history of the panorama. *Early Popular Visual Culture, 6*(3), 219–238.

43. Ivanov, S. S., Ivanov, P. A., & Vitanov, N. V. (2015). Efficient construction of three- and four-qubit quantum gates by global entangling gates. *Physical Review A, 91*(3), 032311.

44. Kilduff, J. E., Karanfil, T., & Weber, W. J. (1996). Competitive interactions among components of humic acids in granular activated carbon adsorption systems: Effects of solution chemistry. *Environmental Science & Technology, 30*(4), 1344–1351.

45. Konjević, N. (1999). Plasma broadening and shifting of non-hydrogenic spectral lines: Present status and applications. *Physics Reports, 316*(6), 339–401.

46. Kornberg, K., Gerlich, S., Haslinger, P., Nimmrichter, S., & Arndt, M. (2012). Colloquium: Quantum interference of clusters and molecules. *Reviews of Modern Physics, 84*(1), 157–173.

47. Kroto, H. W. (1988). The chemistry of the interstellar medium. *Philosophical Transactions of the Royal Society of London. Series A, Mathematical and Physical Sciences, 325*(1587), 405–421.

48. Landau, L. D., & Lifshitz, E. M. (1980). Chapter I–The fundamental principles of statistical physics. In *Course of Theoretical Physics* (Vol. 1980, pp. 1–33).

49. Lavelle, F., Hollywood, L., Caraher, M., McGowan, L., Spence, M., Surgenor, D., & Dean, M. (2017). Increasing intention to cook from basic ingredients: A randomized controlled study. *Appetite, 116*, 502–510.

50. Li, J., Schneider, W. D., Berndt, R., Bryant, O. R., & Crampin, S. (1998). Surface-state lifetime measured by scanning tunneling spectroscopy. *Physical Review Letters, 81*(20), 4464.

51. Li, Q., Hu, R. Q., Qian, Y., & Wu, G. (2012). Cooperative communications for wireless networks: Techniques and applications in LTE-advanced systems. *IEEE Wireless Communications, 19*(2), 22–29.

52. Li, W., Wang, P., Li, J., Wang, L., Gong, S., & Li, D. (2017). Influence of stator parameter variation and phase-shift under synchronizing out of phase on turbine generator electromagnetic field. *IEEE Transactions on Energy Conversion, 32*(2), 525–533.

53. Lindgren, I. (1978). A coupled-cluster approach to the many-body perturbation theory for open-shell systems. *International Journal of Quantum Chemistry, 14*(S12), 33–58.

54. Lorin, E., Chelkowski, S., & Bandrauk, A. (2007). A numerical Maxwell–Schrödinger model for intense laser–matter interaction and propagation. *Computer Physics Communications, 177*(12), 908–932.

55. Machala, Z., Janda, M., Hensel, K., Jedlovský, I., Leštinská, L., Foltin, V., & Morvova, M. (2007). Emission spectroscopy of atmospheric pressure plasmas for bio-medical and environmental applications. *Journal of Molecular Spectroscopy, 243*(2), 194–201.

56. Magonov, S. N. (1993). Surface characterization of materials at ambient conditions by scanning tunneling microscopy (STM) and atomic force microscopy (AFM). *Applied Spectroscopy Reviews, 28*(1–2), 1–121. https://doi.org/10.1080/05704929308018312.

57. Massey, H. S. W., & Moiseiwitsch, B. L. (1951). The application of variational methods to atomic scattering problems-I. The elastic scattering of electrons by hydrogen atoms. *Proceedings of the Royal Society of London. Series A. Mathematical and Physical Sciences, 205*(1083), 483–496. https://doi.org/10.1098/rspa.1951.0171.

58. Mathys, G., & Stenflo, J. O. (1987). Anomalous Zeeman effect and its influence on the line absorption and dispersion coefficients. *Astronomy and Astrophysics, 171*(1–2), 368–377.

59. McCrea, W. H. (1953). Cosmology. *Reports on Progress in Physics, 16*(1), 321. https://doi.org/10.1088/0034-4885/16/1/308.

60. McNamara, D. S. (2001). Reading both high-coherence and low-coherence texts: Effects of text sequence and prior knowledge. *Canadian Journal of Experimental Psychology/Revue canadienne de psychologie expérimentale, 55*(1), 51. https://doi.org/10.1037/h0087354.

61. Meixner, A. J., Renn, A., Bucher, S. E., & Wild, U. P. (1986). Spectral hole burning in glasses and polymer films: The Stark effect. *The Journal of Physical Chemistry, 90*(26), 6777–6785. https://doi.org/10.1021/j100281a015.

62. Morino, Y., & Toyama, M. (1961). Zeeman effect of the nuclear quadrupole resonance spectrum in crystalline powder. *The Journal of Chemical Physics, 35*(4), 1289–1296. https://doi.org/10.1063/1.1732075.

63. Myers, A. B. (1998). Molecular electronic spectral broadening in liquids and glasses. *Annual Review of Physical Chemistry, 49*(1), 267–295. https://doi.org/10.1146/annurev.physchem.49.1.267.

64. Naruse, M., Tate, N., Aono, M., & Ohtsu, M. (2013). Information physics fundamentals of nanophotonic. *Reports on Progress in Physics, 76*(5), 056401. https://doi.org/10.1088/0034-4885/76/5/056401.

65. Patsourakos, S., & Klimchuk, J. A. (2006). Nonthermal spectral line broadening and the nanoflare model. *The Astrophysical Journal, 647*(2), 1452. https://doi.org/10.1086/505491.

66. Pericoli-Ridolfini, V., Giannone, L., & Bartiromo, R. (1994). Frequency spectral broadening of lower hybrid waves in tokamak plasmas-causes and effects. *Nuclear Fusion, 34*(4), 469. https://doi.org/10.1088/0029-5515/34/4/I04.

67. Plantenberg, J. H., De Groot, P. C., Harmans, C. J. P. M., & Mooij, J. E. (2007). Demonstration of controlled-NOT quantum gates on a pair of superconducting quantum bits. *Nature, 447*(7146), 836–839. https://doi.org/10.1038/nature05896.

68. Pokutnyi, S. I., Jacak, L., Misiewicz, J., Salejda, W., & Zegrya, G. G. (2004). Stark effect in semiconductor quantum dots. *Journal of Applied Physics, 96*(2), 1115–1119. https://doi.org/10.1063/1.1767278.

69. Popescu, C. C., Yang, R., Tuffs, R. J., Natale, G., Rushton, M., & Aharonian, F. (2017). A radiation transfer model for the Milky Way: I. Radiation fields and application to high-energy astrophysics. *Monthly Notices of the Royal Astronomical Society, 470*(3), 2539–2558. https://doi.org/10.1093/mnras/stx1407.

70. Pratt, R. H. (2004). Tutorial on fundamentals of radiation physics: Interactions of photons with matter. *Radiation Physics and Chemistry, 70*(4–5), 595–603. https://doi.org/10.1016/j.radphyschem.2004.03.007.

71. Rajagopal, K., & Wilczek, F. (2001). The condensed matter physics of QCD. In M. Shifman (Ed.), *At the Frontier of Particle Physics: Handbook of QCD* (Vol. 3, pp. 2061–2151). World Scientific Publishing. https://doi.org/10.1142/9789812810458_0050.

72. Roueff, E., & Lique, F. (2013). Molecular excitation in the interstellar medium: Recent advances in collisional, radiative, and chemical processes. *Chemical Reviews, 113*(12), 8906–8938. https://doi.org/10.1021/cr400068a.

73. Sadlej, A. J. (1981). Perturbation theory of the electron correlation effects for atomic and molecular properties. *The Journal of Chemical Physics, 75*(1), 320–331. https://doi.org/10.1063/1.442003.

74. Shen, Z., Hou, X., Li, W., Aini, G., Chen, L., & Gong, Y. (2015). Impact of landscape pattern at multiple spatial scales on water quality: A case study in a typical urbanized watershed in China. *Ecological Indicators, 48*, 417–427. https://doi.org/10.1016/j.ecolind.2014.08.010.

75. Simon, D., Stenstrom, D. M., & Read, S. J. (2015). The coherence effect: Blending cold and hot cognitions. *Journal of Personality and Social Psychology, 109*(3), 369–388. https://doi.org/10.1037/pspa0000029.

76. Somsen, O. J., van Grondelle, R., & van Amerongen, H. (1996). Spectral broadening of interacting pigments: Polarized absorption by photosynthetic proteins. *Biophysical Journal, 71*(4), 1934–1951. https://doi.org/10.1016/S0006-3495(96)79488-1.

77. Subramanian, K. S. (1996). Determination of metals in biofluids and tissues: Sample preparation methods for atomic spectroscopic techniques. *Spectrochimica Acta Part B: Atomic Spectroscopy, 51*(3), 291–319. https://doi.org/10.1016/0584-8547(95)01462-0.

78. Sullivan, G. M., & Artino Jr, A. R. (2013). Analyzing and interpreting data from Likert-type scales. *Journal of Graduate Medical Education, 5*(4), 541–542. https://doi.org/10.4300/JGME-5-4-18.

79. Terhal, B. M., Conrad, J., & Vuillot, C. (2020). Towards scalable bosonic quantum error correction. *Quantum Science and Technology, 5*(4), 043001. https://doi.org/10.1088/2058-9565/ab98a5.

80. Uzan, J. P. (2011). Varying constants, gravitation and cosmology. *Living Reviews in Relativity, 14*(1), 1–155. https://doi.org/10.12942/lrr-2011-2.

81. Wang, Z., Huang, B., Xue, R., Huang, X., & Chen, L. (1999). Spectroscopic investigation of interactions among components and ion transport mechanism in polyacrylonitrile based electrolytes. *Solid State Ionics, 121*(1–4), 141–156. https://doi.org/10.1016/S0167-2738(98)00459-7.

82. Werner, J., Wallis, H., & Ertmer, W. (1992). Atoms with anomalous Zeeman effect in a 1D-magneto-optical molasses. *Optics Communications, 94*(6), 525–529. https://doi.org/10.1016/0030-4018(92)90469-V.

83. Winefordner, J. D., Gornushkin, I. B., Correll, T., Gibb, E., Smith, B. W., & Omenetto, N. (2004). Comparing several atomic spectrometric methods to the super stars: Special emphasis on laser induced breakdown spectrometry, LIBS, a future super star. *Journal of Analytical Atomic Spectrometry, 19*(9), 1061–1083. https://doi.org/10.1039/B402871A.

84. Yesilyurt, G., Martin, W. R., & Brown, F. B. (2012). On-the-fly Doppler broadening for Monte Carlo codes. *Nuclear Science and Engineering, 171*(3), 239–257. https://doi.org/10.13182/NSE171-239.

85. Yu, H. G., & Muckerman, J. T. (2002). A general variational algorithm to calculate vibrational energy levels of tetraatomic molecules. *Journal of Molecular Spectroscopy, 214*(1), 11–20. https://doi.org/10.1006/jmsp.2002.8571.

86. Zhao, S., Liu, M., Zhao, L., & Zhu, L. (2018). Influence of interactions among three biomass components on the pyrolysis behavior. *Industrial & Engineering Chemistry Research, 57*(15), 5241–5249. https://doi.org/10.1021/acs.iecr.8b00065.

CHAPTER

2 Atoms

LEARNING OBJECTIVES

After studying this chapter, you will be able to:

1. Understand the energy spectrum and wavefunctions of one-electron atoms and ions;
2. Analyze the concepts of orbital angular momentum and magnetic dipole moment;
3. Explain the significance of the Stern-Gerlach experiment in understanding atomic structure;
4. Describe the effects of nuclear spin and hyperfine structure in atoms;
5. Interpret electronic transitions and selection rules in atomic spectra;
6. Explore the electronic structure of many-electron atoms using the Hartree-Fock method;
7. Apply the fundamentals of spectroscopy to analyze core levels and optical spectra.

KEY TERMS FROM THIS CHAPTER

Angular wavefunction

Hartree-Fock method

Magnetic dipole moment

Radial wavefunction

Energy spectrum

Hyperfine structure

Orbital angular momentum

Stern-Gerlach experiment

UNIT INTRODUCTION

The study of atoms and their interactions is a basis of modern physics and chemistry. Understanding atomic structure and behavior not only provides insight into the fundamental building blocks of matter but also lays the foundation for various applications across science and technology. This chapter explains the details of one-electron and many-electron atoms, exploring their energy levels, wavefunctions, and the effects of external fields on their spectra (Grant & Quiney, 1988).

Chapter begins by examining one-electron atoms and ions, such as hydrogen and hydrogen-like ions, which serve as essential models for more complex atomic systems. These simple systems allow us to introduce key concepts such as energy spectra, angular and radial wavefunctions, orbital angular momentum, and magnetic dipole moments. The Stern-Gerlach experiment and electron spin will also be discussed, highlighting their significance in the development of quantum mechanics. Furthermore, we will explore fine structure and hyperfine structure, which arise from spin-orbit coupling and nuclear interactions, respectively. These phenomena provide deeper insight into the subtleties of atomic energy levels (Pyykkö, 2013).

The latter part of the chapter discusses to many-electron atoms, addressing the complexities introduced by electron-electron interactions. We will discuss identical particles and the independent-particles approximation, which simplify the analysis of multi-electron systems. The section on the two-electron atom will illustrate these principles with specific examples. Advanced computational methods, such as the Hartree-Fock method, will be introduced to describe electronic structure accurately across the periodic table.

Spectroscopy, a powerful tool for probing atomic and molecular structure, is given significant attention. We will cover the fundamentals of spectroscopy, including core levels and spectra, optical spectra, and electric-dipole selection rules. Understanding these concepts is crucial for interpreting experimental data and applying spectroscopic techniques in various scientific fields.

Throughout this chapter, we aim to provide a comprehensive and technical overview of atomic structure and behavior, emphasizing the mathematical and physical principles that underpin these phenomena. The insights gained here will be invaluable for further studies in quantum mechanics, spectroscopy, and related disciplines.

2.1. ONE-ELECTRON ATOM/IONS

In atomic physics, the study of one-electron atoms and ions holds profound significance, offering a window into the foundational principles of quantum mechanics. These systems, consisting of a single electron interacting with a nucleus, represent the quantum nature of matter. Central to this exploration is the characterization of their energy spectrum, which reveals the discrete and quantized nature of energy levels governed by Schrödinger's equation (Mingos & Dahl, 2012). The angular and radial wavefunctions describe the spatial distribution of the electron around the nucleus, providing insights into the probability density of finding the electron at various locations and orientations. The concept of orbital angular momentum and its associated magnetic dipole moment explain the behavior of atoms in external magnetic fields, crucial for applications ranging from magnetic resonance imaging to quantum information processing. Moreover, the Stern-Gerlach experiment demonstrates the quantization of angular momentum and spin, underscoring the intrinsic quantal properties of particles like electrons. Fine structure effects, including relativistic corrections and spin-orbit coupling, further refine our understanding by revealing subtle energy level splitting and spectral line shifts. Collectively, the study of one-electron atoms and ions not only deepens our theoretical grasp of quantum mechanics but also underpins diverse technological innovations across physics and beyond (Salahub, 1987).

2.1.1. The Energy Spectrum

In atomic physics, the energy spectrum of a one-electron atom or ion refers to the discrete set of energy levels that the electron can occupy due to its interaction with the positively charged nucleus. These energy levels are a direct consequence of solving the Schrödinger equation for the system, which describes the quantum state of the electron in terms of its wavefunction, denoted typically by (ψ) (Afrosimov et al., 1986) (Figure 2.1).

The Schrödinger equation for a one-electron atom or ion in its non-relativistic form is:

$$H\,\psi = E\,\psi$$

where; (H) is the Hamiltonian operator, (E) is the energy eigenvalue associated with a particular quantum state, and (ψ) is the corresponding eigenfunction or wavefunction (Lin & Ho, 2015).

The energy spectrum is characterized by quantized energy levels, typically labeled by a principal quantum number (n). The energy levels (E_n) for a hydrogen-like atom (or ion) are given by:

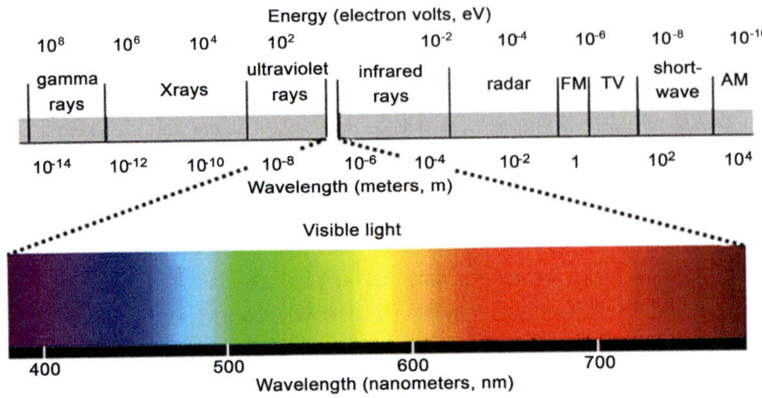

Figure 2.1. *Illustration of the energy spectrum.*

Source: Peter Langdon, Creative Commons License.

Remember

Remember to grasp the foundational concepts of atomic structure discussed in this chapter, as they are crucial for understanding the complexities of atoms and their behavior in various contexts.

$$E_n = -\frac{Z^2 \cdot 13.6\ \text{eV}}{n^2}$$

where; (Z) is the atomic number (number of protons in the nucleus); and (n) is the principal quantum number (taking integer values starting from 1) (Kaneko, 2020).

Each energy level (E_n) corresponds to a specific electronic state with a unique set of quantum numbers that describe the electron's behavior. These quantum numbers include the principal quantum number (n), the azimuthal quantum number (l), and the magnetic quantum number (m_l), which together define the orbital angular momentum and orientation of the electron's orbital.

The energy spectrum reveals important properties such as the energy differences between adjacent levels, which determine the frequencies of light emitted or absorbed during electronic transitions. These transitions are governed by the selection rules, which dictate the allowed changes in quantum numbers $(\Delta n), (\Delta l), (\Delta m_l)$..

Overall, the energy spectrum of one-electron atoms and ions provides a fundamental framework for understanding the discrete nature of atomic energy levels and their implications for spectroscopic observations and technological applications. It underscores the foundational role of quantum mechanics in describing the behavior of matter at atomic scales, essential for both theoretical investigations and practical advancements in fields such as quantum computing and precision measurement (Rahman et al., 2021).

The angular wavefunction in the context of atomic physics refers to the component of the total wavefunction of an electron that describes its angular dependence in spherical coordinates around the nucleus (Misaizu et al., 1992).

2.1.2. The Angular Wavefunction

In spherical coordinates (r, θ, ϕ), where (r) is the radial distance from the nucleus, (θ) is the polar angle measured from the z-axis, and (ϕ) is the azimuthal angle measured from the x-axis in the xy-plane, the angular part of the wavefunction, denoted as $Y_l^m(\theta, \phi)$, is expressed as a spherical harmonic. The spherical harmonics are eigenfunctions of the angular momentum operators (L^2) (total angular momentum squared) and (L_z) (z-component of angular momentum), with quantum numbers (l) (azimuthal quantum number) and (m) (magnetic quantum number) specifying the angular momentum state of the electron (Ajana et al., 2018).

The general form of the angular wavefunction $Y_l^m(\theta, \phi)$ is:

$$Y_l^m(\theta, \phi) = (-1)^m \sqrt{\frac{(2l+1)(l-m)!}{4\pi(l+m)!}} P_l^m(\cos\theta) e^{im\phi}$$

where; $(P_l^m(\cos\theta)$ are the associated Legendre functio(è)ns; $(-1)^m$ is a phase factor; (m) ranges from $(-l)$ to (l) in integer steps; and and (ϕ) are the polar and azimuthal angles, respectively.

The quantum numbers (l) and (m) determine the angular momentum quantum state of the electron (Gutzwiller, 1970):

i. (l) dictates the magnitude of the orbital angular momentum $\left(L = \sqrt{l(l+1)}\right)$;

ii. (m) specifies the projection of the angular momentum along a chosen axis, typically the z-axis.

The angular wavefunctions are orthogonal with respect to the angles (è) and (ϕ), fulfilling the orthonormality condition (Lifshitz, 1964):

$$\int_0^{2\pi} \int_0^{\pi} Y_{l'}^{m'*}(\theta, \phi) Y_l^m(\theta, \phi) \sin\theta \, d\theta \, d\phi = \delta_{l'l}\delta_{m'm}$$

where; $(\delta_{l'l})$and $(\delta_{m'm})$are the Kronecker delta functions.

The angular wavefunctions $Y_l^m(\theta, \phi)$ play a crucial role in determining the selection rules for atomic transitions, as transitions between different angular momentum states are governed by conservation rules that depend on these quantum numbers. For instance, transitions in atoms or ions obey the selection rule $(\Delta l = \pm 1)$ and $(\Delta m = 0, \pm 1)$ (Gor'Kov, 1958).

In conclusion, the angular wavefunction $Y_l^m(\theta, \phi)$ provides a detailed description of the spatial orientation of the electron's orbital angular momentum in an atom or ion. Its mathematical formulation and properties are fundamental in understanding and predicting atomic spectra, magnetic properties, and the angular distribution of electron clouds around nuclei, contributing significantly to the foundational principles of quantum mechanics in atomic physics (Gildener & Patrascioiu, 1977).

2.1.3. The Radial Wavefunction

The radial wavefunction in atomic physics represents the part of the total wavefunction of an electron that depends on the distance from the nucleus, describing how the probability density of finding the electron varies with distance. It plays a crucial role in determining the energy levels and spatial distribution of electrons within atoms or ions (Thompson, 1968).

In the context of solving the Schrödinger equation for a hydrogen-like atom (an atom with only one electron, such as hydrogen or singly ionized helium), the total wavefunction $\Psi(r,\theta,\phi)$ can be separated into radial and angular components:

$$\Psi(r,\theta,\phi) = R_{nl}(r)Y_l^m(\theta,\phi)$$

where; $R_{nl}(r)$ is the radial wavefunction; $Y_l^m(\theta,\phi)$ is the angular wavefunction discussed previously; (n) is the principal quantum number, determining the overall size and energy of the orbital; and (l) is the azimuthal quantum number, related to the magnitude of the orbital angular momentum.

The radial wavefunction $R_{nl}(r)$ satisfies the radial part of the Schrödinger equation and is specific to each quantum state characterized by (n) and (l). It is normalized such that the total probability of finding the electron over all space is unity (Löwdin, 1964):

$$\int_0^\infty |R_{nl}(r)|^2 r^2 dr = 1$$

The radial wavefunction $R_{nl}(r)$ can be expressed in terms of the associated Laguerre polynomials $(L_{n-l-1}^{2l+1}(2r/\alpha))$, where (α) is the Bohr radius and $(L_{n-l-1}^{2l+1}))$ are the generalized Laguerre polynomials. The explicit form of $(R_{nl}(r))$ is given by (McKoy, 1965):

$$R_{nl}(r) = \sqrt{\left(\frac{2}{na}\right)^3 \frac{(n-l-1)!}{2n[(n+l)!]^3}} e^{-r/na} \left(\frac{2r}{na}\right)^l L_{n-l-1}^{2l+1}\left(\frac{2r}{na}\right)$$

where; (a) is the Bohr radius; $(n=1,2,3,\ldots)$ is the principal quantum number; and $(l=0,1,2,\ldots,n-1)$ is the azimuthal quantum number (Jack et al., 2008).

The radial wavefunction $(R_{nl}(r))$ determines several key properties of the electron in the atom:

1. **Energy Levels:** The energy of the electron in a hydrogen-like atom depends primarily on the principal quantum number (n), with higher (n) corresponding to higher energy levels.
2. **Probability Density:** The square of the radial wavefunction $\left(|R_{nl}(r)|^2\right)$ gives the probability density of finding the electron at a distance (r) from the nucleus. This probability

density exhibits nodes and peaks that reflect the quantized nature of the electron's allowed energy states.

3. **Ionization Probability:** The behavior of $R_{nl}(r)$ near large (r) (far from the nucleus) dictates the probability of ionization, where the electron is no longer bound to the nucleus due to its energy exceeding the ionization energy (Anderson & Raithel, 2012).

Moreover, the radial wavefunction $R_{nl}(r)$ is instrumental in understanding spectroscopic transitions and the fine structure of atomic spectra. The discrete energy levels of atoms and ions observed experimentally are directly related to the solutions of the Schrödinger equation, with $R_{nl}(r)$ providing insights into the spatial distribution of electrons in these quantum systems (Shrivastava, 2011).

2.1.4. Orbital Angular Momentum and Magnetic Dipole Moment

Orbital angular momentum $(\vec{L})$ is a fundamental concept in quantum mechanics that describes the rotational motion of a particle around a fixed point, such as an electron orbiting a nucleus in an atom. In the context of atomic physics, it is quantized and intimately related to the quantum numbers (l) (azimuthal quantum number) and (m) (magnetic quantum number), which characterize the spatial distribution and orientation of the electron's wavefunction (Figure 2.2.).

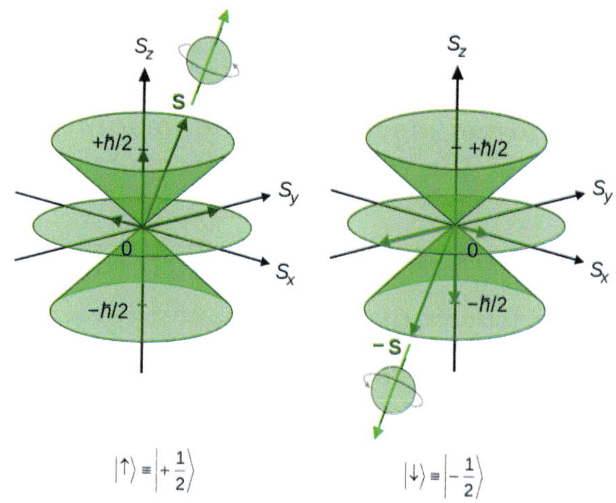

Figure 2.2. *Illustration of the orbital angular momentum.*

Source: Leon Drennan, Creative Commons License.

2.1.4.1. Angular Momentum Operators

The orbital angular momentum operator in quantum mechanics is defined as:

$$\vec{L} = \vec{r} \times \vec{p}$$

where; $(\vec{r})$ is the position vector of the electron relative to the nucleus; and $(\vec{p})$ is its momentum operator. For a single electron in a hydrogen-like atom, the components of the orbital angular momentum operator in spherical coordinates are (Cowan & Griffin, 1976):

$$L_z = -i\hbar \frac{\partial}{\partial \phi}$$

$$\vec{L^2} = -\hbar^2 \left[\frac{1}{\sin\theta} \frac{\partial}{\partial\theta} \left(\sin\theta \frac{\partial}{\partial\theta} \right) + \frac{1}{\sin^2\theta} \frac{\partial^2}{\partial\phi^2} \right]$$

These operators satisfy the commutation relations:

$$[L_i, L_j] = i\hbar \epsilon_{ijk} L_k$$

2.1.4.2. Quantization of Angular Momentum

In quantum mechanics, the magnitude squared of the orbital angular momentum $\left(\vec{L^2}\right)$ and its component along a specified axis $\left(L_z\right)$ have quantized eigenvalues:

$$\vec{L^2}Y_{lm}(\theta,\phi) = \hbar^2 l(l+1)Y_{lm}(\theta,\phi)$$

$$L_z Y_{lm}(\theta,\phi) = \hbar m Y_{lm}(\theta,\phi)$$

where; (l) is the azimuthal quantum number (taking integer values $(0,1,2,\ldots,n-1)$); and (m) is the magnetic quantum number (taking values $(-l,-l+1,\ldots,l-1,l)$) (Durand, 1986).

2.1.5. Magnetic Dipole Moment

The magnetic dipole moment $\left(\vec{\mu}\right)$ of an electron due to its orbital motion is given by:

$$\vec{\mu} = -\gamma\vec{L}$$

where; (γ) is the gyromagnetic ratio and depends on the properties of the electron. The z-component of the magnetic dipole moment $\left(\mu_z\right)$ can be expressed as:

$$\mu_z = -\gamma L_z$$

2.1.5.1. Importance and Applications

Understanding orbital angular momentum and magnetic dipole moments is crucial for several aspects of atomic physics (Salvat et al., 1995):

1. **Fine Structure of Atomic Spectra:** The quantization of $(\vec{L})$ leads to the splitting of energy levels into fine structure components, observed in atomic spectra due to interactions like spin-orbit coupling.
2. **Stern-Gerlach Experiment:** The measurement of electron spin and orbital angular momentum played a pivotal role in the development of quantum mechanics, as demonstrated by the Stern-Gerlach experiment.
3. **Magnetic Properties:** The magnetic dipole moment contributes to the magnetic properties of atoms and molecules, influencing their behavior in external magnetic fields and contributing to phenomena such as Zeeman splitting.
4. **Quantum Information Processing:** Orbital angular momentum and magnetic dipole moments are also important in the context of quantum information processing and quantum computing, where the manipulation of quantum states relies on these intrinsic properties (Newton, 1960).

2.1.6. The Stern-Gerlach Experiment

The Stern-Gerlach experiment, conducted by Otto Stern and Walther Gerlach in 1922, was a groundbreaking experiment that provided definitive evidence for the quantization of angular momentum and the discrete nature of spin in quantum mechanics. The experiment involved passing a beam of silver atoms through an inhomogeneous magnetic field and observing their deflection onto a detector screen. This deflection revealed that the silver atoms split into discrete beams rather than forming a continuous distribution, demonstrating that angular momentum (in this case, the electron's spin) is quantized (Xayavong & Smirnova, 2022) (Figure 2.3).

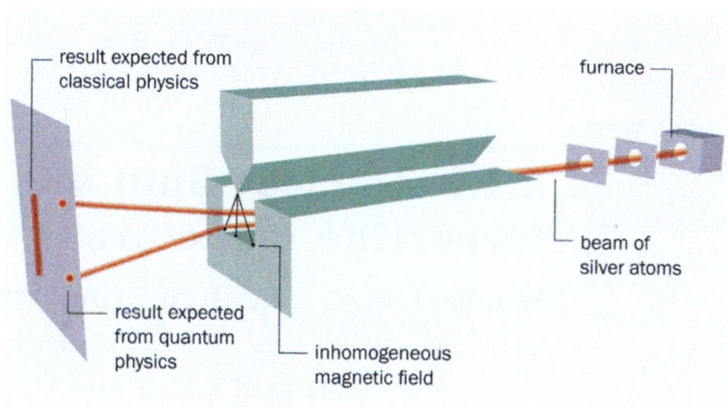

Figure 2.3. *Setup of the Stern-Gerlach experiment.*

Source: Physics World, Creative Commons License.

In the Stern-Gerlach experiment:

1. **Beam of Silver Atoms:** A beam of silver atoms is produced by heating a silver source, which then passes through a collimator to produce a narrow, well-defined beam.
2. **Inhomogeneous Magnetic Field:** The beam of silver atoms enters an inhomogeneous magnetic field created by a magnet with a strong gradient. The magnetic field gradient is crucial because it causes a spatially varying magnetic force on the atoms.
3. **Observation Screen:** After passing through the magnetic field, the beam of silver atoms is directed onto a detector screen or photographic plate. This screen records the positions where the atoms strike after passing through the magnetic field (Armstrong, 1970).

Did you know?

Did you know that the Stern-Gerlach experiment demonstrated the quantization of angular momentum in atomic systems? This pivotal experiment provided evidence for the existence of quantized angular momentum states in atoms.

2.1.6.1. Quantum Mechanical Explanation

According to quantum mechanics, the magnetic moment $\left(\vec{i}\right)$ of an electron (and thus the silver atom) due to its spin interacts with the magnetic field $\left(\vec{B}\right)$:

$$\vec{\tau} = -\vec{\mu} \times \vec{B}$$

The potential energy associated with this interaction is given by:

$$U = -\vec{\mu} \cdot \vec{B}$$

where; $(\vec{\tau})$ is the torque experienced by the magnetic moment $(\vec{\mu})$

In the Stern-Gerlach experiment, the spatially varying magnetic field exerts a force on the magnetic moment of the electrons, causing the atoms to separate into distinct beams rather than forming a continuous distribution. This observed splitting is due to the quantization of the electron's magnetic moment component along the direction of the magnetic field gradient (Guo et al., 2020).

2.1.6.2. Results and Implications

The Stern-Gerlach experiment provided several key insights and implications for quantum mechanics (Mansuripur, 2011):

1. **Quantization of Angular Momentum:** The discrete splitting of the silver atom beam demonstrated that angular momentum (specifically spin angular momentum) is quantized. This was a pivotal confirmation of the postulates of quantum mechanics.
2. **Spin Quantization:** The experiment provided direct evidence for the quantization of spin angular momentum, revealing that electrons possess intrinsic angular momentum that can only take certain discrete values.
3. **Foundation of Quantum Theory:** The experiment played a crucial role in establishing the foundations of quantum theory, demonstrating that physical observables such as angular momentum are quantized and can only take specific discrete values.
4. **Technological Applications:** The principles and techniques developed in the Stern-Gerlach experiment have practical applications in fields such as quantum computing, magnetic resonance imaging (MRI), and precision measurements where the manipulation and measurement of spin states are essential (Grillo, 2017).

2.1.7. Nuclear Spin and Hyperfine Structure

Nuclear spin and hyperfine structure play critical roles in understanding the fine details of atomic spectra and the interactions between atomic nuclei and electrons. This section explores the quantum mechanical properties of nuclear spin, the resultant hyperfine interactions, and their implications for atomic spectroscopy (Steinmetz et al., 2019).

2.1.7.1. Nuclear Spin

Nuclear spin refers to the intrinsic angular momentum possessed by atomic nuclei due to the presence of protons and neutrons. This property is analogous to electron spin but arises from different quantum mechanical principles. The nuclear spin (I) is quantized and characterized by the quantum number (I), which can take on integer or half-integer values depending on the number of protons and neutrons in the nucleus (Larocque et al., 2016) (Figure 2.4).

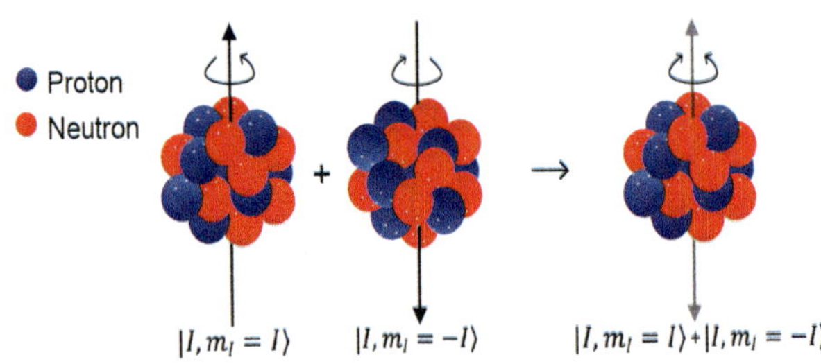

Figure 2.4. *Nuclear spin Schrödinger cat state: a superposition |I, I + |I, −I (ignoring the normalization constant) of spin-up and spin-down orientations of a high-spin nucleus.*

Source: Pragati Gupta, Creative Commons License.

The magnetic moment $(\overrightarrow{\mu_I})$ associated with nuclear spin interacts with external magnetic fields $(\vec{B})$, giving rise to energy levels that depend on the orientation of $(\overrightarrow{\mu_I})$ relative to $(\vec{B})$. The energy levels are given by:

$$E_I = -\overrightarrow{\mu_I} \cdot \vec{B} = -\gamma_I \hbar I_z B$$

where; (γ_I) is the gyromagnetic ratio of the nucleus; $(\hbar)$ is the reduced Planck's constant; (I_z) is the z-component of the nuclear spin quantum number (I); and (B) is the external magnetic field (Inamori et al., 2011).

2.1.7.2. Hyperfine Structure

Hyperfine structure refers to the splitting of atomic energy levels caused by the interaction between the magnetic moments of the atomic nucleus (nuclear spin) and the electrons' orbital and spin magnetic moments. This interaction is often much weaker than the interactions involving the electron's spin and orbital angular momenta but is crucial for resolving fine details in atomic spectra (Inamori et al., 2011) (Figure 2.5).

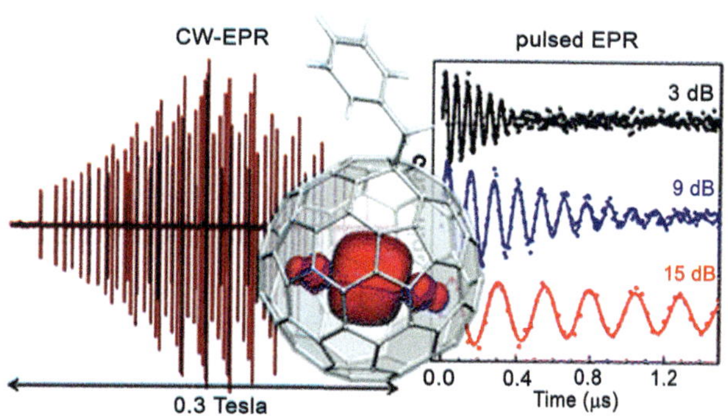

Figure 2.5. *Large hyperfine structure of the electron spin levels in an endohedral dimetallofullerene and its spin coherent properties.*

Source: Ruslan B. Zaripov, Creative Commons License.

The Hamiltonian describing the hyperfine interaction (H_{hf}) can be expressed as:

$$H_{\text{hf}} = A\vec{I} \cdot \vec{J}$$

where; (A) is the hyperfine coupling constant; $(\vec{I})$ is the nuclear spin operator; and $(\vec{J})$ is the total electronic angular momentum operator (sum of orbital and spin contributions) (Lewis, 1994).

2.1.7.3. Zeeman Effect and Hyperfine Splitting

In the presence of an external magnetic field ($\vec{B}$), the Zeeman effect (Figure 2.6) leads to the splitting of hyperfine levels into multiple sublevels, each corresponding to different projections of the total angular momentum $\left(\vec{F} = \vec{I} + \vec{J}\right)$ along the magnetic field axis:

$$E_{F,m_F} = g_F \mu_B B m_F$$

where; (g_F) is the Landé g-factor for the hyperfine structure; (μ_B) is the Bohr magneton; (B) is the magnetic field strength; and (m_F) is the magnetic quantum number associated with the total angular momentum (F) (Bohr, 1951).

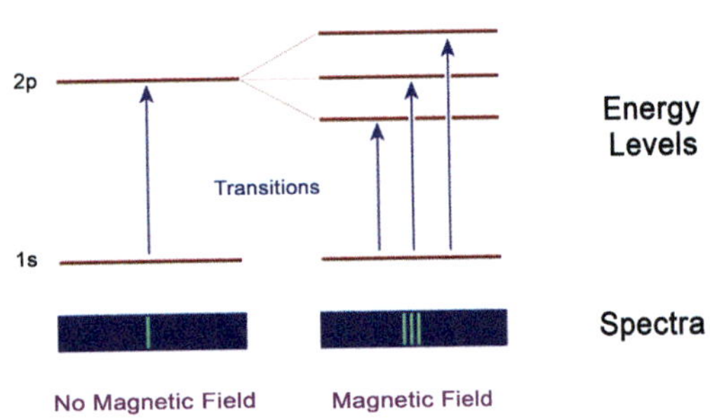

Figure 2.6. *Schematic diagram of the Zeeman effect.*

Source: Pulkit Chandel, Creative Commons License.

- **Experimental Observations:** Experimental techniques such as microwave spectroscopy and optical spectroscopy are employed to observe hyperfine structure in atomic spectra. Microwave transitions occur between hyperfine levels, providing precise measurements of hyperfine splitting and allowing determination of the nuclear spin quantum numbers (I) for different isotopes (Eisinger & Jaccarino, 1958).

The study of hyperfine structure has numerous practical applications:

1. **Atomic Clocks:** The hyperfine transitions in atoms such as cesium-133 are used as the basis for atomic clocks, which are essential for precise timekeeping and global positioning systems (GPS).
2. **Nuclear Magnetic Resonance (NMR):** NMR spectroscopy relies on the principles of nuclear spin and hyperfine interactions to study the structure and dynamics of molecules in chemistry, biology, and medicine.
3. **Quantum Computing:** Hyperfine qubits formed by atomic nuclei with long coherence times are potential candidates for quantum information processing.

In conclusion, nuclear spin and hyperfine structure are fundamental concepts in atomic physics, providing insights into the quantum mechanical properties of atomic nuclei and their interactions with electrons. The detailed understanding of hyperfine interactions has paved the way for technological advancements in precision measurement and quantum technologies (Bohr & Weisskopf, 1950).

2.1.8. Electronic Transitions and Selection Rules

Electronic transitions and their selection rules govern the absorption and emission of electromagnetic radiation by atoms, playing a pivotal role in spectroscopy and understanding atomic structure within quantum mechanics. These transitions occur when an electron moves from one energy level (orbitals) to another within an atom, accompanied by the absorption or emission of photons whose energy corresponds to the energy difference between the initial and final states of the electron (White, 1930).

2.1.8.1. Energy Levels and Transitions

In quantum mechanics, an electron in an atom occupies discrete energy levels described by quantum numbers. The energy levels are determined by the solutions to the Schrödinger equation for the atom's potential energy, typically the Coulomb potential between the nucleus and the electron (Figure 2.7). The energy of an electron in the nth energy level is given by (Abragam & Pryce, 1951):

$$E_n = -\frac{Z^2 R_H}{n^2}$$

where; (Z) is the atomic number; (R_H) is the Rydberg constant; and (n) is the principal quantum number.

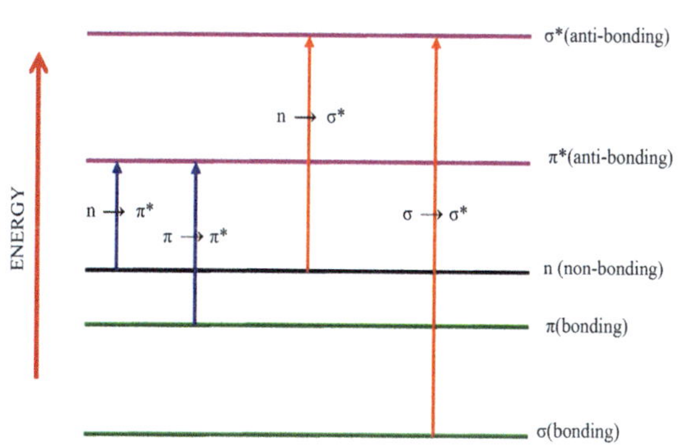

Figure 2.7. *Energy level diagrams with electronic transitions.*

Source: Priya Murugasen, Creative Commons License.

- **Absorption and Emission of Photons:** Electronic transitions occur when an electron absorbs a photon of energy (ΔE), causing it to move from a lower-energy (initial) state to a higher-energy (final) state. The energy of the absorbed photon must exactly match the energy difference between the initial and final states (Lloyd & Pake, 1954):

$$\Delta E = E_{\text{final}} - E_{\text{initial}}$$

Conversely, when an electron transitions from a higher-energy state to a lower-energy state, it emits a photon with an energy (ΔE):

$$\Delta E = E_{\text{initial}} - E_{\text{final}}$$

2.1.8.2. Selection Rules

Selection rules dictate which electronic transitions are allowed or forbidden based on quantum mechanical principles. These rules are derived from conservation laws such as conservation of angular momentum and conservation of parity. The selection rules specify:

1. **Change in Quantum Numbers:** The principal quantum number (n), azimuthal quantum number (l), and magnetic quantum number (m) can change during an electronic transition, subject to certain restrictions.

- **Angular Momentum Conservation:** The total angular momentum (J) and its projection $(\quad)$ must be conserved. The total angular momentum (J) is the vector sum of the orbital angular momentum (L) and the spin angular momentum (S):

$\vec{J} = \vec{L} + \vec{S}$

2. **The selection rule** $(\Delta J = 0, \pm 1)$**:** It indicates that transitions where $(\Delta J \neq 0)$ are typically weaker or forbidden (Petersen & Shugart, 1962).
3. **Parity Conservation:** Parity is a quantum number that describes the symmetry properties of the wavefunction. The selection rule $(\Delta l = \pm 1)$ indicates that transitions where the parity changes are often weaker or forbidden.

Remember

When analyzing spectra involving electric-dipole transitions, remember that the selection rules dictate which transitions are allowed based on changes in the quantum numbers of the electronic states involved.

2.1.9. Spectra in a Magnetic Field

Spectra in a magnetic field, also known as Zeeman effect, is a phenomenon where the spectral lines of atoms split and polarize in the presence of a magnetic field. This effect was discovered by Pieter Zeeman in 1896 and played a crucial role in the development of quantum mechanics, providing evidence for the quantization of angular momentum.

2.1.9.1. Zeeman Effect Basics

In the absence of a magnetic field, atoms exhibit discrete spectral lines corresponding to transitions between energy levels dictated by quantum mechanics. However, when a magnetic field $(\vec{B})$ is applied, the energy levels of atoms split into multiple components due to the interaction between the magnetic moment of the atom and the magnetic field. This splitting is known as the Zeeman effect (Bleaney, 1951).

2.1.9.2. Interaction Energy

The interaction energy (E_{int}) between the magnetic moment $(\vec{\mu})$ of an atom and the magnetic field $(\vec{B})$ is given by:

$$E_{\text{int}} = -\vec{\mu} \cdot \vec{B}$$

where; $(\vec{\mu})$ is the magnetic moment of the atom. For an electron, the magnetic moment $(\vec{\mu})$ is related to its spin $(\vec{S})$ and orbital angular momentum $(\vec{L})$ by:

$\vec{\mu} = -g\mu_B\vec{S} - g\mu_B\vec{L}$

where; (g) is the g-factor (Landé g-factor); $(\grave{\imath}_B)$ is the Bohr magneton; and $(\vec{S})$ and $(\vec{L})$ are the spin and orbital angular momentum vectors, respectively (Goudsmit & Gropper, 1931).

2.1.9.3. Classification of Zeeman Effect

Zeeman effect is classified into three types based on the relative orientations of the magnetic field $(\vec{B})$, the angular momentum $(\vec{J})$, and the external magnetic field:

1. **Normal Zeeman Effect:** Occurs when the magnetic field $(\vec{B})$ is weak, and the spectral lines split into three components: (π) (unshifted) and $(\sigma_{\pm})\,(splitby(\pm\Delta E))$.
2. **Anomalous Zeeman Effect:** Occurs when the magnetic field $(\vec{B})$ is strong enough to cause additional splitting of spectral lines due to interactions beyond the linear approximation. It is more complex and involves higher-order corrections.
3. **Paschen-Back Effect:** Occurs when the magnetic field $(\vec{B})$ is very strong, leading to the merging of spectral lines into broader groups. This effect is observed when the energy separation between (ΔE) and $(\Delta E_{\mathrm{HFS}})$is smaller than $(g\mu_B B)$

2.1.9.4. Mathematical Formulation

The Zeeman effect can be mathematically described using perturbation theory in quantum mechanics. The perturbed energy levels $(E_{n,m})$ for a given electron state (n) and magnetic quantum number (m) are:

$E_{n,m} = E_n + g\mu_B Bm$

where; (E_n) is the unperturbed energy level; (g) is the Landé g-factor; (μ_B) is the Bohr magneton; (B) is the magnetic field strength; and (m) is the magnetic quantum number (Callis et al., 1983).

2.2. MANY ELECTRONS ATOMS

In the study of atomic physics, Many-Electron Atoms involve examining how multiple electrons interact within an atom. Unlike simpler models of single-electron atoms, where each electron's behavior can be studied independently, many-electron systems require more complex theoretical frameworks. This section explores concepts such as electron correlations, which describe how interactions between electrons affect their energies and spatial arrangements. Techniques like the Hartree-Fock method are used to approximate the behavior of these electrons, considering factors like mutual repulsion and their impact on atomic structure and spectra. Understanding many-electron atoms is essential for applications in fields like materials science and quantum chemistry, providing insights into phenomena such as chemical bonding and the electronic properties of solids (Satten, 1957).

2.2.1. Identical Particles

In quantum mechanics, identical particles are particles that cannot be distinguished from each other even in principle. This indistinguishability has profound implications for the quantum description of systems with multiple particles, such as atoms with multiple electrons (Figure 2.8).

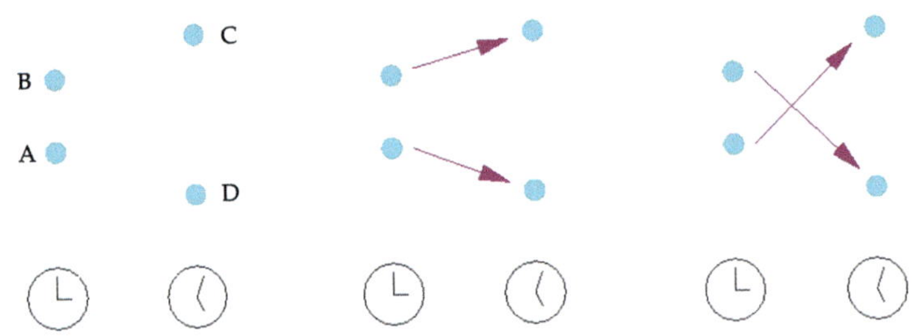

Figure 2.8. *Indistinguishable particles in quantum mechanics: (left) initially there are two particles at A and B, later two particles are found at C and D; (middle) but we cannot be certain whether the particles travelled from A to D and B to C or (right) from A to C and B to D, because they are identical.*

Source: TCM Group, Creative Commons License.

2.2.1.1. Indistinguishability and Symmetry

Identical particles, such as electrons in an atom, bosons in a Bose-Einstein condensate, or fermions in a degenerate electron gas, obey the principles of quantum statistics. For fermions, described by the Pauli exclusion principle, no two identical fermions can occupy the same quantum state simultaneously. This leads to the formation of electron shells in atoms and governs the behavior of electrons in solids.

2.2.1.2. Mathematical Formulation

Mathematically, the quantum state of a system of identical particles is described by a wavefunction that must be symmetric for bosons (particles with integer spin) and antisymmetric for fermions (particles with half-integer spin). For two fermions, the antisymmetric wavefunction under particle exchange is given by (Gruetter et al., 1998):

$$\Psi_{\text{antisymmetric}}(1,2) = \frac{1}{\sqrt{2}}\left(\psi_1(1)\psi_2(2) - \psi_1(2)\psi_2(1)\right)$$

where; (ψ_1) and (ψ_2) represent the wavefunctions of the two fermions, and the terms inside the parentheses represent the spatial wavefunctions of each fermion.

2.2.1.3. Applications in Atomic Physics

Understanding identical particles is crucial in atomic physics for predicting and interpreting atomic spectra, particularly in many-electron atoms where electron-electron interactions can significantly affect the energy levels and transitions. This knowledge is foundational in spectroscopy, where the symmetry of wavefunctions determines the selection rules governing allowed transitions between energy levels.

2.2.1.4. Experimental Verification

Experimental verification of quantum statistics and the indistinguishability of particles is demonstrated in experiments such as the Stern-Gerlach experiment for spin-1/2 particles, where the outcome depends on the quantum state's symmetry. Furthermore, measurements in condensed matter physics confirm the formation of energy bands and gaps, crucial for understanding electronic behavior in materials (Kulsrud & Anderson, 1992).

2.2.2. The Independent-Particles

The independent-particles approximation is a fundamental approach in quantum mechanics used to simplify the description of many-electron systems, such as atoms and molecules, where interactions between particles are complex and challenging to solve exactly. This approximation assumes that the behavior of each particle within the system can be effectively described independently of the others, neglecting direct interactions between particles except through an averaged or effective potential (Lites et al., 1991).

2.2.2.1. Mathematical Foundation

In the framework of quantum mechanics, the wavefunction (Ψ) describing a system of (N) particles can be written as a product of single particle wavefunctions:

$$\Psi(r_1, r_2, \ldots, r_N) = \psi_1(r_1)\psi_2(r_2)\ldots\psi_N(r_N)$$

where; $(\psi_i(r_i))$ represents the wavefunction of the (i)-th particle in the system. This factorization assumes that the particles are distinguishable and that their wavefunctions do not depend on the positions of other particles. While this approach simplifies the mathematical treatment of the system, it does not account for the correlations and interactions that may exist between particles.

2.2.2.2. Hartree Approximation

A common implementation of the independent-particles approximation is the Hartree approximation, which accounts for the average interaction between particles through an effective potential. In the Hartree approximation, each particle moves in the mean field generated by the others. The Hamiltonian of the system is then (Ledbetter et al., 2009):

$$\hat{H} = \sum_{i=1}^{N}\left[-\frac{\hbar^2}{2m}\nabla_i^2 + V_{\text{ext}}(r_i)\right] + \sum_{i=1}^{N} V_{\text{eff}}(r_i)$$

where; $V_{\text{ext}}(r_i)$ is the external potential acting on the i-th particle; and $(V_{\text{eff}}(r))$ is the effective potential arising from the average interaction with all other particles.

2.2.2.3. Applications and Limitations

The independent-particles approximation is particularly useful in atomic and molecular physics to calculate energy levels and wavefunctions of many-electron systems. It forms the basis for the Hartree-Fock method, which iteratively determines self-consistent single-particle wavefunctions considering the average interaction. However, this method neglects correlation effects between particles, such as electron-electron interactions beyond the mean field (Sinanoğlu, 1962).

2.2.2.4. Beyond Hartree-Fock

To address the shortcomings of the independent-particles approximation, more advanced methods like configuration interaction (CI), coupled cluster (CC), and density functional theory (DFT) are employed. These methods incorporate electron correlation effects more rigorously by considering multiple-particle interactions and are crucial for accurately predicting properties of molecules and condensed matter systems (Sinanoğlu, 1961).

2.2.3. The Two-Electron Atom

The study of two-electron atoms within quantum mechanics serves as a crucial steppingstone towards understanding more complex atomic systems and molecules. Unlike the simpler one-electron atoms, such as hydrogen or helium ions, which can be solved analytically using the Schrödinger equation, the two-electron system introduces significant challenges due to the electron-electron interaction. This interaction leads to a more intricate energy spectrum, wavefunction structure, and spectroscopic properties (Sucher, 1980).

- **Schrödinger Equation for Two-Electron Atoms:** The Hamiltonian $(\hat{H})$ for a two-electron atom in atomic units (where $(\hbar=1)$ and $(m\_e=1)$) is given by:

$$\hat{H} = -\frac{1}{2}(\nabla_1^2 + \nabla_2^2) - \frac{Z}{r_1} - \frac{Z}{r_2} + \frac{1}{r_{12}}$$

where; (∇_i^2) represents the Laplacian operator for the (i)-th electron; (r_i) denotes the distance of the (i)-th electron from the nucleus of charge (Z); and (r_{12}) is the distance between the two electrons.

2.2.3.1. Separation of Variables

Due to the complexity of the electron-electron interaction term $\left(\frac{1}{r_{12}}\right)$, the Schrödinger equation cannot be solved analytically in closed form for more than one electron. However, by applying the method of separation of variables and assuming a product form for the wavefunction

$$\Psi(r_1, r_2) = \psi_1(r_1)\psi_2(r_2)$$

where; $(\psi_i(r_i))$ represents the wavefunction of the ith electron, we can simplify the problem into two coupled equations:

$$\left[-\frac{1}{2}\nabla_1^2 - \frac{Z}{r_1}\right]\psi_1(r_1) + \int \frac{\psi_2(r_2)}{|r_1 - r_2|}dr_2 = E\psi_1(r_1)$$

$$\left[-\frac{1}{2}\nabla_2^2 - \frac{Z}{r_2}\right]\psi_2(r_2) + \int \frac{\psi_1(r_1)}{|r_1 - r_2|}dr_1 = E\psi_2(r_2)$$

where E which is sume of $E\psi_1(r_1)$; $E\psi_2(r_2)$ represents the total energy of the two-electron system. Solving these coupled equations requires advanced numerical methods or approximation techniques due to the presence of the electron-electron repulsion term.

2.2.3.2. Energy Levels and Spectroscopic Features

The energy spectrum of two-electron atoms exhibits several notable features compared to one-electron systems. Instead of discrete energy levels, the spectrum shows a series of closely spaced energy states due to the coupling between the electrons. This coupling results in energy splitting known as the fine structure, which arises from the interaction of orbital angular momenta, electron spins, and the electron-nucleus potential (Sinanoğlu, 1961).

2.2.3.3. Approximation Methods

One approach to solve the Schrödinger equation for two-electron atoms is the variational method, where trial wavefunctions are employed to approximate the ground and excited state energies. Another method, known as the perturbation theory, involves treating the electron-electron interaction as a small perturbation to the Hamiltonian of the non-interacting (one-electron) atom (Glass & Hibbert, 1978).

2.2.4. The Hartree-Fock Method

The Hartree-Fock (HF) method is a pivotal approach in quantum mechanics used to approximate the wavefunction of a many-electron system, such as atoms and molecules. It seeks to improve upon the independent-electron approximation by incorporating the effects of electron-electron interactions through a self-consistent procedure. This method forms the foundation for understanding electronic structure and properties in chemistry, physics, and materials science.

2.2.4.1. Theoretical Basis

The HF method begins with the premise that the wavefunction of a many-electron system can be approximated as a product of single-electron wavefunctions, known as spin orbitals. The HF wavefunction $(\O_{HF})$ is constructed as (Lennard, 1931):

$$\Psi_{HF}(r_1, r_2, \ldots, r_N) = \psi_1(r_1)\psi_2(r_2)\cdots\psi_N(r_N)$$

where; $(\psi_i(r_i))$ are the single electron wavefunctions (spin orbitals) for electrons $(i = 1, 2, \ldots, N)$.

2.2.4.2. Hartree-Fock Equations

The goal of the HF method is to minimize the total energy (E_{HF}) of the system by solving the HF equations, derived from the variation of the total energy with respect to the spin orbitals under the constraints of orthonormality:

$$\widehat{f}_i \psi_i(r_i) = \epsilon_i \psi_i(r_i)$$

where; $(\widehat{f}_i)$ is the Fock operator for electron (i); which includes both kinetic energy and the average potential created by the other electrons:

$$\widehat{f}_i = -\frac{1}{2}\nabla_i^2 + \sum_{j=1}^{N}\left[\frac{Z}{|r_i - R_j|}\right] + \int \frac{\rho(r_2, r_3, \ldots, r_N)}{|r_1 - r_2|} dr_2 dr_3 \ldots dr_N$$

where; $(\rho(r_2, r_3, \ldots, r_N))$ is the electron density excluding electron (r_1); and (ϵ_i) denotes the orbital energy of electron (i).

2.2.4.3. Self-Consistent Field Approach

The HF method iteratively determines the spin orbitals $(\psi_i(r_i))$ by solving the HF equations until self-consistency is achieved. The procedure involves the following steps:

1. **Initial Guess:** Start with an initial set of guess spin orbitals.
2. **Construct Density:** Calculate the electron density $(\rho(r))$ from the occupied spin orbitals.
3. **Solve Fock Equations:** Use the current density to construct the Fock operator and solve the HF equations for each spin orbital.
4. **Update Orbitals:** Update the spin orbitals based on the solutions to the HF equations.
5. **Iterate:** Repeat the process until convergence criteria, such as small changes in the total energy or spin orbitals, are met.

2.2.4.4. Advantages and Limitations

The HF method provides a reasonable approximation to the true many electron wavefunction and yields insights into the electronic structure, ionization energies, and molecular geometries of atoms and molecules. However, it has limitations due to neglecting dynamic electron correlation effects beyond the mean-field approximation. Post-Hartree-Fock methods, such as Configuration Interaction (CI) and Coupled Cluster (CC) theory, aim to correct these deficiencies by including electron correlation effects explicitly (Sapirstein, 1993).

2.2.4.5. Applications

The HF method finds widespread application in computational chemistry and quantum chemistry software packages. It is used to predict molecular properties, simulate chemical reactions, and design new materials. Its development has paved the way for more sophisticated electronic structure methods, contributing significantly to the fields of theoretical chemistry and materials science (Figure 2.9).

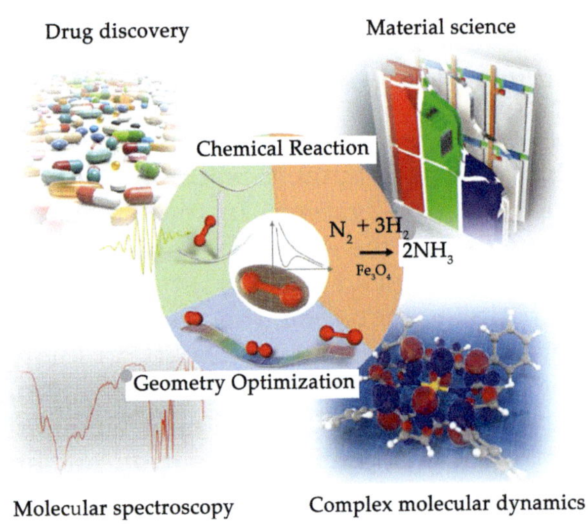

Figure 2.9. *Various applications of Hartree-Fock method.*

2.2.5. Electronic Structure Across the Periodic Table

Electronic structure across the periodic table refers to the arrangement and behavior of electrons in atoms as characterized by their quantum states, energy levels, and configurations. This topic is crucial for understanding the chemical and physical properties of elements and compounds, playing a fundamental role in fields ranging from chemistry and materials science to astrophysics. The organization of electrons in atoms is governed by quantum mechanics, with each element uniquely defined by its atomic number (Z), which determines the number of protons in its nucleus and, therefore, the number of electrons in its neutral state (Coleman et al., 2007).

2.2.5.1. Quantum Mechanical Basis

In quantum mechanics, the electronic structure of an atom is described by the Schrödinger equation, which, for the non-relativistic case, governs the behavior of electrons in the electrostatic potential created by the atomic nucleus:

$$\hat{H}\Psi(r_1, r_2, \dots, r_N) = E\Psi(r_1, r_2, \dots, r_N)$$

where; $(\hat{H})$ is the Hamiltonian operator; (Ψ) is the wavefunction of the electrons; (r_i) denotes the position coordinates of electron (i); and (E) is the total energy of the system.

2.2.5.2. Atomic Orbitals and Energy Levels

The solution to the Schrödinger equation yields discrete energy levels characterized by quantum numbers. The principal quantum number (n) determines the main energy level or shell, while the azimuthal quantum number (l), magnetic quantum number (m_l), and spin quantum number (m_s) further specify the orbital characteristics. The energy levels within each shell degenerate for a hydrogen-like atom but are split by the presence of multiple electrons due to electron-electron interactions.

$$E_{nl} = -\frac{Z^2 R_\infty}{n^2}$$

where; (R_∞) is the Rydberg constant and (Z) is the atomic number. This equation describes the energy levels of an electron in a hydrogen-like atom, which is modified for multi-electron atoms to include the effective nuclear charge and electron-electron interactions.

2.2.5.3. Periodicity and the Periodic Table

The periodic table organizes elements based on their atomic number and chemical properties, reflecting the recurring patterns in electronic structure as electrons fill successive shells and subshells. The periodic table is divided into periods (rows) and groups (columns), with each period representing a new energy level being filled and each group sharing similar electronic configurations and chemical behaviors (Shabaev et al., 2013).

2.2.5.4. Trends in Electronic Structure

Several trends in electronic structure emerge across the periodic table:

1. **Ionization Energy:** The energy required to remove an electron from

an atom increases across a period due to increased nuclear charge and decreasing atomic radius.

2. **Electron Affinity:** The energy released when an electron is added to an atom becomes more negative (more favorable) from left to right across a period.
3. **Atomic Radius:** Decreases across a period due to increasing effective nuclear charge attracting electrons closer to the nucleus.
4. **Electronegativity:** Increases across a period due to increased nuclear charge, making atoms more likely to attract electrons in chemical bonding.

2.2.5.5. Complex Electronic Configurations

As one moves across the periodic table, elements exhibit complex electronic configurations involving filled and partially filled shells and subshells, leading to variations in chemical reactivity and physical properties. Transition metals, for instance, exhibit multiple oxidation states due to the availability of electrons in d-orbitals, contributing to their catalytic and magnetic properties (Srivastava et al., 2018).

2.2.5.6. Computational Methods and Beyond

Computational methods, including the Hartree-Fock method and density functional theory (DFT), are employed to calculate electronic structures and predict properties of atoms and molecules across the periodic table. These methods account for electron-electron interactions and relativistic effects, providing insights into bonding, spectroscopy, and material properties (Lemmen & Lengauer, 2000).

2.2.6. Fundamentals of Spectroscopy

Fundamentals of spectroscopy encompass the principles and techniques used to study the interaction of matter with electromagnetic radiation. Spectroscopy plays a pivotal role in modern physics, chemistry, astronomy, and materials science by providing detailed information about the structure, composition, and dynamics of atoms, molecules, and materials

2.2.6.1. Electromagnetic Radiation and Transition Processes

Spectroscopy revolves around the absorption, emission, or scattering of electromagnetic radiation by matter. Electromagnetic radiation, characterized by its wavelength (λ) or frequency (ν) interacts with atoms and molecules through transitions between quantum states. The energy (E) of a photon is related to its frequency by $(E = h\nu)$, where (h) is Planck's constant. The selection rules govern which transitions are allowed based on changes in quantum numbers such as (Δn), (Δl), and $(\Delta m_l)$for electronic transitions (Sliwoski et al., 2014).

$$I \propto N_2 - N_1$$

2.2.6.2. Types of Spectroscopic Techniques

1. **UV-Visible Spectroscopy:** Involves light in the ultraviolet and visible regions, probing electronic transitions such as $(\pi)-(\pi^*)$ and $(n)-(\pi^*)$ transitions in organic molecules.

2. **Infrared Spectroscopy (IR):** Probes molecular vibrations by measuring absorption or emission of infrared radiation. It reveals characteristic vibrational modes corresponding to bonds in molecules.
3. **Raman Spectroscopy:** Measures inelastic scattering of photons, providing information about rotational and vibrational transitions in molecules. Raman shifts correspond to energy differences between initial and final states.

2.2.6.3. Advanced Techniques and Applications

Modern spectroscopic techniques incorporate advanced methods such as Fourier-transform spectroscopy, which enhances spectral resolution and sensitivity. Time-resolved spectroscopy explores dynamic processes by measuring changes in spectra over time, crucial for understanding chemical reactions and molecular dynamics. Applications span diverse fields, including environmental monitoring, pharmaceutical analysis, and astronomical observations of distant galaxies and stars (Filippenko, 1997).

2.2.6.4. Computational Spectroscopy

Computational methods, including quantum chemical calculations and molecular dynamics simulations, complement experimental spectroscopy. Density functional theory (DFT) and ab initio methods predict electronic spectra and molecular properties, guiding experimental design and interpretation.

2.2.7. Core Levels and Spectra

Core levels and spectra in spectroscopy refer to the energy levels associated with tightly bound inner-shell electrons in atoms or molecules. These inner-shell electrons, also known as core electrons, exhibit characteristic binding energies and spectral features that are distinct from those of valence electrons. This section explores the theoretical underpinnings and experimental techniques used to study core levels and their spectra, which are crucial for understanding electronic structure and chemical bonding.

2.2.7.1. Core Level Binding Energies

The binding energy (E_{core}) of a core electron in an atom or molecule is the energy required to remove the electron from its inner-shell orbital. This binding energy is influenced by the nuclear charge (Z), shielding effects from other electrons, and relativistic effects. The binding energy can be estimated using empirical relationships or calculated using quantum mechanical methods such as Hartree-Fock theory or Density Functional Theory (DFT). For an atom with nuclear charge (Z), the binding energy (E_{core}) can be approximated by the empirical formula (Phillips, 1966):

$$E_{\text{core}} = -13.6 \cdot Z^2 \cdot \left(\frac{1}{n_{\text{core}}}\right)^2 \quad \text{eV}$$

where; (n_{core}) is the principal quantum number of the core electron.

2.2.7.2. X-ray Photoelectron Spectroscopy (XPS)

X-ray photoelectron spectroscopy is a powerful experimental technique used to measure core-level binding energies. In XPS, a sample is irradiated with X-rays, which ionize core electrons by ejecting them from their respective orbitals. The energy of the ejected electron is measured, yielding a spectrum that reveals the binding energies of different core levels. The core-level binding energies in XPS are influenced by chemical environment, oxidation states, and electronic structure variations.

2.2.7.3. Auger Electron Spectroscopy (AES)

Auger electron spectroscopy probes core-level transitions by detecting Auger electrons emitted when an atom relaxes after core-level ionization. The Auger electron energy spectrum provides information about the energy levels of core electrons and the structure of atoms or molecules. Auger transitions are governed by energy conservation rules and selection rules like those in XPS.

2.2.7.4. Chemical Shifts and Spectral Interpretation

Core-level binding energies exhibit chemical shifts due to differences in chemical environment and electronic structure. These shifts, often measured in electron volts (eV), are sensitive to changes in oxidation states, bonding configurations, and coordination environments of atoms or molecules. The chemical shifts provide valuable insights into chemical bonding, surface chemistry, and electronic interactions within materials.

2.2.7.5. Applications in Materials Science and Chemistry

Core-level spectroscopy finds applications in diverse fields such as catalysis, surface science, and nanotechnology. For instance, XPS and AES are essential tools for characterizing catalyst surfaces, identifying surface contaminants, and studying electronic properties of nanomaterials. Core-level spectra serve as fingerprints for elemental analysis and quantitative determination of chemical composition in complex materials.

2.2.7.6. Future Directions and Advanced Techniques

Advancements in synchrotron radiation sources and ultrafast spectroscopy techniques continue to enhance the resolution, sensitivity, and time resolution of core-level spectroscopy. Time-resolved X-ray techniques offer unprecedented insights into ultrafast electronic dynamics and chemical reactions. Computational methods such as DFT and many-body perturbation theory complement experimental spectra by predicting core-level energies and interpreting complex spectral features.

In conclusion, core levels and spectra play a crucial role in elucidating electronic structure and chemical properties at the atomic and molecular scales. From fundamental studies of inner-shell electrons to practical applications in materials science, core-level spectroscopy remains indispensable for advancing our understanding of matter and its interactions in various scientific disciplines.

2.2.8. Optical Spectra

Optical spectra refer to the electromagnetic radiation absorbed or emitted by atoms, molecules, or solids in the optical (visible) region of the electromagnetic spectrum.

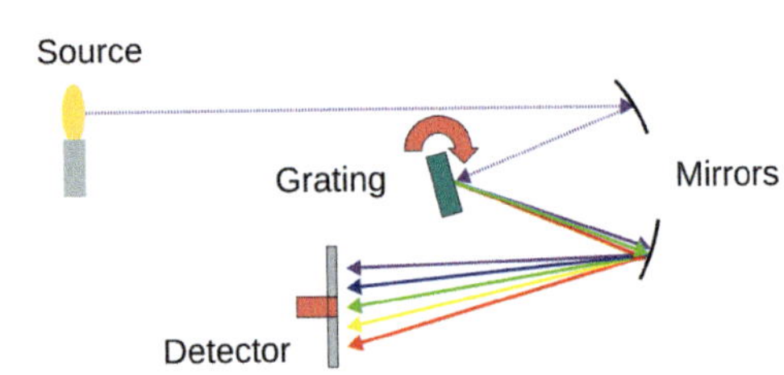

Figure 2.10. *Schematic of optical spectroscopy.*

Source: Jooja, Creative Commons License.

This section delves into the theoretical foundations, experimental techniques, and applications of optical spectra, encompassing absorption, emission, and scattering phenomena (Giangrande et al., 1980) (Figure 2.10).

2.2.8.1. Fundamentals of Optical Spectroscopy

Optical spectroscopy involves the interaction of electromagnetic radiation with matter, where photons of specific energies (corresponding to wavelengths in the visible range) interact with electronic, vibrational, or rotational states of the system. The energy (E) of a photon is related to its wavelength (λ) by the equation:

$$E = \frac{hc}{\lambda}$$

where; (h) is Planck's constant and (c) is the speed of light in vacuum. The absorption or emission of photons leads to transitions between energy levels in the system, resulting in characteristic spectral features.

2.2.8.2. Absorption Spectroscopy

Absorption spectroscopy measures the absorption of light by a sample as a function of photon energy (or wavelength). The absorption spectrum provides information about the electronic structure and energy levels of the material. The Beer-Lambert law describes the relationship between the absorption coefficient (α), concentration (c), path length (L), and incident intensity (I_0):

$$I = I_0 e^{-\alpha c L}$$

where; (I) is the transmitted intensity. Optical absorption spectra are often recorded using UV-visible spectrophotometers, which measure the amount of light absorbed by a sample across a range of wavelengths (Knight & Milonni, 1980).

2.2.8.3. Emission Spectroscopy

Emission spectroscopy investigates the light emitted by a sample upon excitation, providing insights into electronic transitions and energy states. Fluorescence and phosphorescence are examples of emission processes where excited electrons return to lower energy states, emitting photons of specific energies. The emission spectrum reflects the energy levels and electronic transitions within the material.

2.2.8.4. Raman Spectroscopy

Raman spectroscopy involves the inelastic scattering of photons by molecular vibrations or rotational transitions within a sample. The Raman effect results in shifts in the energy (frequency) of scattered photons, providing information about molecular vibrations and crystal structures. The Raman shift $(\Delta\omega)$ is related to the vibrational energy (ω) of the molecule:

$$\Delta\omega = \omega_{\text{incident}} - \omega_{\text{scattered}}$$

2.2.8.5. Applications of Optical Spectroscopy

Optical spectroscopy finds widespread applications in various fields, including chemistry, physics, materials science, and biology. In chemistry, UV-visible absorption spectroscopy is used for quantitative analysis, determining reaction kinetics, and studying electronic transitions in transition metal complexes. In materials science, optical spectra reveal band gaps in semiconductors, optical properties of nanomaterials, and defects in crystals. In biology, fluorescence spectroscopy aids in studying biomolecular structures and interactions.

2.2.8.6. Advanced Techniques and Future Directions

Advanced optical spectroscopy techniques include time-resolved spectroscopy, where ultrafast lasers probe transient states and dynamics of molecular processes. Fourier-transform spectroscopy improves spectral resolution and sensitivity by utilizing interferometric techniques. Nonlinear optical spectroscopy explores higher-order light-matter interactions, offering insights into nonlinear optical properties and quantum phenomena.

2.2.9. Electric-Dipole Selection Rules

Electric-dipole selection rules in spectroscopy govern the allowed transitions between quantum states of atoms or molecules due to the interaction with an electric field. These rules arise from the conservation of angular momentum and parity considerations and play a crucial role in determining which transitions are observable in optical and infrared spectra. This section delves into the theoretical underpinnings of electric-dipole transitions, their mathematical formulation, and their implications in spectroscopic analysis (Knight & Milonni,1980).

2.2.9.1. Theoretical Basis

The interaction between an electric dipole moment $(\vec{d})$ and an electric field $(\vec{E})$ gives rise to an energy change (ΔE) for an atom or molecule:

$$\Delta E = -\vec{d} \cdot \vec{E}$$

In quantum mechanics, the transition probability (P) between initial state (ψ_i) and final state (ψ_f) due to an electric dipole interaction is governed by the transition dipole moment integral $(\langle\psi_f|\vec{d}|\psi_i\rangle)$:

$$P \propto |\langle\psi_f|\vec{d}|\psi_i\rangle|^2$$

The matrix element $(\langle\psi_f|\vec{d}|\psi_i\rangle)$ encapsulates the quantum mechanical probability amplitude for the transition. The selection rules emerge from the conditions under which this integral is non-zero, ensuring conservation of angular momentum and parity.

2.2.9.2. Selection Rules

Electric-dipole transitions typically involve changes in the angular momentum quantum number $(\Delta l = \pm 1)$. This means that for an atom or molecule to undergo an observable transition, the initial and final states must differ by one unit of angular momentum. For example, an electron transitioning between two different orbitals within an atom will typically change its orbital angular momentum by one unit.

2.2.9.3. Parity Considerations

Parity conservation requires that the parity of the initial and final states must be the same for an electric-dipole transition to be allowed. The parity of a wavefunction (ψ) is determined by its behavior under spatial inversion $(\vec{r} \rightarrow -\vec{r})$ States with even parity (symmetric under inversion) can only transition to other states with even parity, and similarly for states with odd parity.

2.2.9.4. Mathematical Formulation

The transition dipole moment ($\langle\psi_f|\vec{d}|\psi_i\rangle$) can be expressed using the position operator $(\vec{r})$:

$$_f|\vec{d}|\psi_i\rangle = \int \psi_f^*(\vec{r})\vec{r}\,\psi_i(\vec{r})\, d$$

where; (ψ_f) and (ψ_i) are the final and initial wavefunctions, respectively. This integral must be non-zero for the transition to be allowed, subject to the selection rules discussed.

2.2.9.5. Applications

Electric-dipole selection rules are fundamental in interpreting experimental spectroscopic data. For instance, in optical spectroscopy, transitions between electronic energy levels in atoms or molecules are governed by these rules, allowing scientists to identify and characterize atomic and molecular spectra. Understanding these rules also facilitates the design of lasers, where specific transitions are exploited to achieve stimulated emission (MacDonald, 1955).

ACTIVITY 2.1.

Explore the fundamentals of spectroscopy by examining the emission spectra of different elements using a spectroscope. Observe and analyze the unique spectral lines to understand how atomic structure influences the wavelengths of emitted light.

SUMMARY

- Chapter 2 focuses on the fundamental structure and behavior of atoms, with an emphasis on both one-electron atoms/ions and many-electron atoms. It begins by exploring the energy spectrum and wavefunctions, particularly the angular and radial components that describe electron behavior and properties within an atom. The concept of orbital angular momentum and its associated magnetic dipole moment is also discussed, along with the important findings of the Stern-Gerlach experiment, which demonstrated the quantization of angular momentum.
- The chapter further examines nuclear spin and hyperfine structure, detailing how these factors influence atomic spectra. It also covers electronic transitions and selection rules, which govern the emission and absorption of light by atoms. For many-electron atoms, the Hartree-Fock method is introduced as a key computational approach for approximating electronic structures. Additionally, the chapter analyzes the electronic structure across the periodic table, core levels, and optical spectra, highlighting the role of electric-dipole selection rules.

REVIEW QUESTIONS

1. What is the significance of the energy spectrum in one-electron atoms?
2. How do the angular and radial wavefunctions differ in their description of atomic orbitals?
3. Explain the concept of orbital angular momentum and its relation to magnetic dipole moment.
4. What were the key findings of the Stern-Gerlach experiment, and why are they important?
5. Describe the role of nuclear spin in determining the hyperfine structure of an atom.
6. How do electronic transitions and selection rules influence atomic spectra?
7. What is the Hartree-Fock method, and how does it contribute to our understanding of many-electron atoms?

MULTIPLE CHOICE QUESTIONS

1. Which experiment is crucial in demonstrating the quantization of angular momentum?

a. Rutherford's gold foil experiment

b. Stern-Gerlach experiment

c. Millikan oil-drop experiment

d. Davisson-Germer experiment

2. **In the context of atomic physics, what does hyperfine structure refer to?**
   a. The splitting of spectral lines due to electron spin
   b. The splitting of spectral lines due to nuclear spin
   c. The broadening of spectral lines due to temperature
   d. The shifting of spectral lines due to external magnetic fields
3. **Which method is used to approximate the electronic structure of many-electron atoms?**
   a. Schrödinger equation
   b. Hartree-Fock method
   c. Heisenberg uncertainty principle
   d. Born-Oppenheimer approximation
4. **What is the primary cause of spectral line splitting in the presence of a magnetic field?**
   a. Doppler effect
   b. Zeeman effect
   c. Stark effect
   d. Compton effect
5. **Which principle states that no two electrons can occupy the same quantum state simultaneously?**
   a. Hund's rule
   b. Aufbau principle
   c. Pauli exclusion principle
   d. Heisenberg uncertainty principle
6. **In a one-electron atom, what determines the shape of an electron's orbital?**
   a. Energy spectrum
   b. Radial wavefunction
   c. Angular wavefunction
   d. Magnetic dipole moment
7. **What is the main purpose of the Stern-Gerlach experiment in atomic physics?**
   a. To measure the charge of an electron
   b. To demonstrate the wave nature of electrons
   c. To observe the quantization of angular momentum
   d. To determine the mass of an electron

Answers to Multiple Questions

1. (b); 2. (b); 3. (b); 4. (b); 5. (c); 6. (c); 7. (c).

REFERENCES

1. Abragam, A., & Pryce, M. H. L. (1951). Theory of the nuclear hyperfine structure of paramagnetic resonance spectra in crystals. *Proceedings of the Royal Society of London. Series A. Mathematical and Physical Sciences, 205*(1080), 135–153.
2. Afrosimov, V. V., Basalaev, A. A., Panov, M. N., & Samo'ilov, A. V. (1986). Electron capture from helium atoms into various electronic states by multiply charged argon ions. *Zh. Eksp. Teor. Fiz, 91*, 476.
3. Ajana, I., Khalil, D., & Makhoute, A. (2018). Signatures of the atomic nucleus in laser-assisted single ionization of one-electron atoms. *The European Physical Journal D, 72*, 1–7.
4. Anderson, S. E., & Raithel, G. (2012). Dependence of Rydberg-atom optical lattices on the angular wave function. *Physical Review Letters, 109*(2), 2–10.
5. Armstrong, L. L. (1970). Group properties of radial wavefunctions. *Le Journal de Physique Colloques, 31*(C4), C4–17.
6. Bleaney, B. (1951). L. Hyperfine structure in paramagnetic salts and nuclear alignment. *The London, Edinburgh, and Dublin Philosophical Magazine and Journal of Science, 42*(328), 441–458.
7. Bohr, A. (1951). Nuclear magnetic moments and atomic hyperfine structure. *Physical Review, 81*(3), 331.
8. Bohr, A., & Weisskopf, V. F. (1950). The influence of nuclear structure on the hyperfine structure of heavy elements. *Physical Review, 77*(1), 94.
9. Bonin, K. D., & McIlrath, T. J. (1984). Two-photon electric-dipole selection rules. *JOSA B, 1*(1), 52–55.
10. Callis, P. R., Scott, T. W., & Albrecht, A. C. (1983). Perturbation selection rules for multiphoton electronic spectroscopy of neutral alternant hydrocarbons. *The Journal of Chemical Physics, 78*(1), 16–22.
11. Coleman, A. J., Rosina, M., Mazziotti, D. A., Erdahl, R. M., Braams, B. J., Percus, J. K., & Kais, S. (2007). Reduced-density-matrix mechanics: With application to many-electron atoms and molecules. *International Journal of Quantum Chemistry, 134*, 1–9.
12. Cowan, R. D., & Griffin, D. C. (1976). Approximate relativistic corrections to atomic radial wave functions. *JOSA, 66*(10), 1010–1014.
13. Durand, B., & Durand, L. (1986). Improved WKB radial wave functions in several bases. *Physical Review A, 33*(5), 28–87.
14. Eisinger, J., & Jaccarino, V. (1958). Effects of the distribution of nuclear magnetization on hyperfine structure. *Reviews of Modern Physics, 30*(2), 528.
15. Filippenko, A. V. (1997). Optical spectra of supernovae. *Annual Review of Astronomy and Astrophysics, 35*(1), 309–355.
16. Giangrande, A., Giovannelli, F., Bartolini, C., Guarnieri, A., & Piccioni, A. (1980). Optical spectra of HDE 245770 = A 0535+ 26. *Astronomy and Astrophysics Supplement Series, 40*, 289–294.

17. Gildener, E., & Patrascioiu, A. (1977). Pseudoparticle contributions to the energy spectrum of a one-dimensional system. *Physical Review D, 16*(2), 423.

18. Glass, R., & Hibbert, A. (1978). Relativistic effects in many-electron atoms. *Computer Physics Communications, 16*(1), 19–34.

19. Gor'Kov, L. P. (1958). On the energy spectrum of superconductors. *Sov. Phys. JETP,* 7(505), 158.

20. Goudsmit, S., & Gropper, L. (1931). Many-electron selection rules. *Physical Review, 38*(2), 225.

21. Grant, I. P., & Quiney, H. M. (1988). Foundations of the relativistic theory of atomic and molecular structure. In *Advances in Atomic and Molecular Physics* (Vol. 23, pp. 37–86).

22. Grillo, V., Tavabi, A. H., Venturi, F., Larocque, H., Balboni, R., Gazzadi, G. C., & Karimi, E. (2017). Measuring the orbital angular momentum spectrum of an electron beam. *Nature Communications, 8*(1), 15–53.

23. Gruetter, R., Weisdorf, S. A., Rajanayagan, V., Terpstra, M., Merkle, H., Truwit, C. L., & Ugurbil, K. (1998). Resolution improvements in *in vivo* 1H NMR spectra with increased magnetic field strength. *Journal of Magnetic Resonance, 135*(1), 260–264.

24. Guo, Y., Zhu, G., Bian, W., Dong, B., & Fang, Y. (2020). Orbital angular momentum dichroism caused by the interaction of electric and magnetic dipole moments and the geometrical asymmetry of chiral metal nanoparticles. *Physical Review A, 102*(3), 033–052.

25. Gutzwiller, M. C. (1970). Energy spectrum according to classical mechanics. *Journal of Mathematical Physics, 11*(6), 1791–1806.

26. Inamori, T., Sako, N., & Nakasuka, S. (2011). Magnetic dipole moment estimation and compensation for an accurate attitude control in nano-satellite missions. *Acta Astronautica, 68*(11–12), 2038–2046.

27. Jack, B., Padgett, M. J., & Franke-Arnold, S. (2008). Angular diffraction. *New Journal of Physics, 10*(10), 10–13.

28. Kaneko, T., Usami, R., Morioka, H., Saitoh, Y., Chiba, A., & Narumi, K. (2020). Electron-loss and destruction processes in collision of MeV/atom carbon cluster ions with rare gases. *Nuclear Instruments and Methods in Physics Research Section B: Beam Interactions with Materials and Atoms, 478*, 218–223.

29. Knight, P. L., & Milonni, P. W. (1980). The Rabi frequency in optical spectra. *Physics Reports, 66*(2), 21–107.

30. Kulsrud, R. M., & Anderson, S. W. (1992). The spectrum of random magnetic fields in the mean field dynamo theory of the galactic magnetic field. *Astrophysical Journal, Part 1, 396*(2), 606–630.

31. Larocque, H., Bouchard, F., Grillo, V., Sit, A., Frabboni, S., Dunin-Borkowski, R. E., & Karimi, E. (2016). Nondestructive measurement of orbital angular momentum for an electron beam. *Physical Review Letters, 117*(15), 54–80.

32. Ledbetter, M. P., Crawford, C. W., Pines, A., Wemmer, D. E., Knappe, S., Kitching, J., & Budker, D. (2009). Optical detection of NMR J-spectra at zero magnetic field. *Journal of Magnetic Resonance, 199*(1), 25–29.

33. Lemmen, C., & Lengauer, T. (2000). Computational methods for the structural alignment of molecules. *Journal of Computer-Aided Molecular Design, 14*, 215–232.

34. Lennard-Jones, J. E. (1931). Wave functions of many-electron atoms. *Mathematical Proceedings of the Cambridge Philosophical Society, 27*(3), 469–480.

35. Lewis, R. R. (1994). Anapole moment of a diatomic polar molecule. *Physical Review A, 49*(5), 3376.

36. Lifshitz, I. M. (1964). The energy spectrum of disordered systems. *Advances in Physics, 13*(52), 483–536.

37. Lin, C. H., & Ho, Y. K. (2015). Shannon information entropy in position space for two-electron atomic systems. *Chemical Physics Letters, 633*, 261–264.

38. Lites, B. W., Bida, T. A., Johannesson, A., & Scharmer, G. B. (1991). High-resolution spectra of solar magnetic features. II—Magnetic fields of umbral brightenings. *Astrophysical Journal, Part 1, 373*, 683–694.

39. Lloyd, J. P., & Pake, G. E. (1954). Spin relaxation in free radical solutions exhibiting hyperfine structure. *Physical Review, 94*(3), 579.

40. Löwdin, P. O. (1964). Angular momentum wavefunctions constructed by projector operators. *Reviews of Modern Physics, 36*(4), 966.

41. MacDonald, W. M. (1955). Isotopic spin selection rule for electric dipole transitions. *Physical Review, 98*(1), 60.

42. Mansuripur, M. (2011). Spin and orbital angular momenta of electromagnetic waves in free space. *Physical Review A—Atomic, Molecular, and Optical Physics, 84*(3), 033–083.

43. McIver Jr, J. W., & Stanton, R. E. (1972). Symmetry selection rules for transition states. *Journal of the American Chemical Society, 94*(24), 8618–8620.

44. McKoy, V. (1965). Angular-Momentum Conditions for a Correlated Wavefunction. *The Journal of Chemical Physics, 43*(5), 1605–1610.

45. Mingos, D. M. P., Day, P., & Dahl, J. P. (2012). *Molecular Electronic Structures of Transition Metal Complexes II* (Vol. 143).

46. Misaizu, F., Sanekata, M., Tsukamoto, K., Fuke, K., & Iwata, S. (1992). Photodissociation of size-selected aquamagnesium (Mg+(H_2O) n) ions for n=1 and 2. *The Journal of Physical Chemistry, 96*(21), 8259–8264.

47. Newton, R. G. (1960). Analytic properties of radial wave functions. *Journal of Mathematical Physics, 1*(4), 319–347.

48. Park, S. R., Kim, C. H., Yu, J., Han, J. H., & Kim, C. (2011). Orbital-angular-momentum based origin of Rashba-type surface band splitting. *Physical Review Letters, 107*(15), 156803.

49. Petersen, F. R., & Shugart, H. A. (1962). Nuclear spin, hyperfine structure, and nuclear moments of 6.8-day lutetium-177. *Physical Review, 126*(1), 252.

50. Phillips, J. C. (1966). The fundamental optical spectra of solids. *Solid State Physics, 18*, 55–164.

51. Pyykkö, P. (2013). *Relativistic Theory of Atoms and Molecules III: A Bibliography 1993–1999* (Vol. 76).

52. Salahub, D. R. (1987). Transition-Metal Atoms and Dimers. *Advances in Chemical Physics: Ab Initio Methods in Quantum Chemistry Part 2, 69*, 447–520.

53. Salvat, F., Fernández-Varea, J. M., & Williamson Jr, W. (1995). Accurate numerical solution of the radial Schrödinger and Dirac wave equations. *Computer Physics Communications, 90*(1), 151–168.

54. Sapirstein, J. (1993). Theory of many-electron atoms. *Physica Scripta, 1993*(T46), 52.

55. Satten, R. A. (1957). Crystalline field selection rules: The effect of vibration-electronic interaction. *The Journal of Chemical Physics, 27*(1), 286–299.

56. Shabaev, V. M., Tupitsyn, I. I., & Yerokhin, V. A. (2013). Model operator approach to the Lamb shift calculations in relativistic many-electron atoms. *Physical Review A—Atomic, Molecular, and Optical Physics, 88*(1), 012–051.

57. Shrivastava, K. N. (2011). Laughlin's Wave Function and Angular Momentum. *International Journal of Modern Physics B, 25*(10), 1301–1357.

58. Sinanoğlu, O. (1961). Many-electron theory of atoms and molecules. *Proceedings of the National Academy of Sciences, 47*(8), 1217–1226.

59. Sinanoğlu, O. (1962). Many-electron theory of atoms and molecules. II. *The Journal of Chemical Physics, 36*(12), 3198–3208.

60. Sliwoski, G., Kothiwale, S., Meiler, J., & Lowe, E. W. (2014). Computational methods in drug discovery. *Pharmacological Reviews, 66*(1), 334–395.

61. Srivastava, A., Nagai, T., Srivastava, A., Miyashita, O., & Tama, F. (2018). Role of computational methods in going beyond X-ray crystallography to explore protein structure and dynamics. *International Journal of Molecular Sciences, 19*(11), 3401.

62. Steinmetz, A., Formanek, M., & Rafelski, J. (2019). Magnetic dipole moment in relativistic quantum mechanics. *The European Physical Journal A, 55*(3), 40.

63. Stodden, V., McNutt, M., Bailey, D. H., Deelman, E., Gil, Y., Hanson, B., & Taufer, M. (2016). Enhancing reproducibility for computational methods. *Science, 354*(6317), 1240–1241.

64. Sucher, J. (1980). Foundations of the relativistic theory of many-electron atoms. *Physical Review A, 22*(2), 348.

65. Thompson, M. W. (1968). The energy spectrum of ejected atoms during the high energy sputtering of gold. *Philosophical Magazine, 18*(152).

66. Ur Rahman, F., Sarwono, Y. P., & Zhang, R. Q. (2021). Solution of two-electron Schrödinger equations using a residual minimization method and one-dimensional basis functions. *AIP Advances, 11*(2).

67. White, H. E. (1930). Nuclear spin and hyperfine structure. *Physical Review, 35*(5), 441.

68. Xayavong, L., & Smirnova, N. A. (2022). Radial overlap correction to superallowed $0+\rightarrow 0+$ nuclear β decays using the shell model with Hartree-Fock radial wave functions. *Physical Review C, 105*(4), 044308.

CHAPTER

3 Molecular Structure and Bonding

LEARNING OBJECTIVES

After studying this chapter, you will be able to:

1. Understand the basic concepts of molecular shape and how it influences chemical properties;
2. Differentiate between various types of isomers and their significance in chemistry;
3. Analyze molecular formulas and their components to understand molecular structure;
4. Comprehend the concept of resonance and its role in stabilizing molecular structures;
5. Explain the nature and types of atomic and molecular orbitals;
6. Describe the process of orbital hybridization and its impact on molecular geometry;
7. Apply VSEPR theory to predict and explain the shapes of molecules based on electron-pair repulsion.

KEY TERMS FROM THIS CHAPTER

Atomic orbitals
Hybrid orbitals
Molecular orbitals
Resonance
Electron-pair repulsion
Isomers
Molecular shape
VSEPR theory

UNIT INTRODUCTION

The notions of molecular structure and bonds are basic in the field of physics, playing a critical role in comprehending the characteristics and behavior of materials. Central to these principles is the notion that atoms unite to create molecules, and the type of bonds between atoms determines the structure of the molecule. The bonding kinds, namely ionic, covalent, metallic, and van der Waals, are determined by how atoms share or transfer electrons (Chiang et al., 2014). Covalent bonds are formed when atoms share electrons, resulting in strong bonds that serve as the structural foundation of organic molecules. Ionic bonds arise from the exchange of electrons between atoms, resulting in the creation of charged ions that mutually attract one another. Metallic bonds consist of a collective of delocalized electrons surrounding a lattice comprising metal ions, whereas van der Waals forces are feeble and temporary interactions between molecules (Shibata et al., 1964).

The significance of molecular structure and bonding encompasses a broad spectrum of physical properties and events. For example, the ability of materials to carry current can be explained by their bonding properties: metals, which have delocalized electrons, exhibit high electrical conductivity, while insulators with covalent bonds do not. The boiling point and melting point of substances are determined by the strength of their bonds. Ionic and covalent compounds generally have greater melting points than substances held together through weaker van der Waals forces. Molecular structure plays a crucial role in determining the shape and arrangement of molecules, which subsequently impact their reactivity, optical characteristics, and connections to other molecules (Haaland, 1979).

In the field of materials engineering and science, a thorough comprehension of molecular structure and bonding is crucial to creating and producing novel materials that possess specific qualities as needed. For instance, the advancement of ceramics, polymers, and modern composites depends on the manipulation of molecular bonds and structure to attain precise thermal, mechanical, and electrical properties (Birk & Kurtz, 1999). The functionality of macromolecules, such as DNA and proteins, in biology is intrinsically linked to their molecular makeup, which impacts several aspects ranging from the activity of enzymes to genetic encoding. Hence, the examination of molecular structure and bonding not only offers an understanding of the underlying characteristics of substances but also propels technological progress in several fields (Richmond, 2001).

3.1. MOLECULAR SHAPE

The molecular structure, which refers to the three-dimensional shape or arrangement of a molecule, is a crucial attribute. This form is determined by the preferred spatial arrangement of covalent bonds to atoms that have a minimum of two bonding partners. Models are the most effective tool for visualizing three-dimensional arrangements. To visually depict these arrangements on a flat surface like paper, a chalkboard, or a screen, perspective drawings are commonly employed. These drawings indicate the direction of a chemical connection by using lines that connect the atoms involved in the interaction. Typically, the configuration mostly revolves around a carbon atom, and the lines indicating bond directions will start from there. The illustration on the right defines a straight line as a bond that lies roughly in the surface plane. The two bonds connected to substitutes A in the left structure are of this type. A wedge-shaped bond is oriented in front of this plane, with the thick end facing the observer, as indicated by the bond to substituent B. The subsequent illustrations employ this nomenclature and highlight the significance of incorporating non-bonding valence shell pair electrons (shown in blue) while examining such combinations (Hillyer & Gibb, 2016).

The prediction of bonding configurations can be easily determined using valence-shell electron-pair repulsive theory, popularly known as VSEPR in introductory chemistry literature. This basic model is founded on the principle that electrons exhibit repulsion towards one another, and it is logical to anticipate that the bonds and non-bonding valence pairs of electrons linked to a certain atom will strive to maintain maximum distance from each other. The bonding arrangements of carbon can be easily memorized, as they can be classified into only three groups (Grant et al., 1996) (Table 3.1).

Table 3.1. *Tabular Data of Different Molecular Shapes*

Configuration	Bonding Partners	Bond Angles	Example
Tetrahedral config	04	109.5°	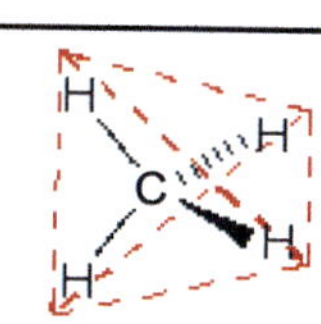
Trigonal config	03	120°	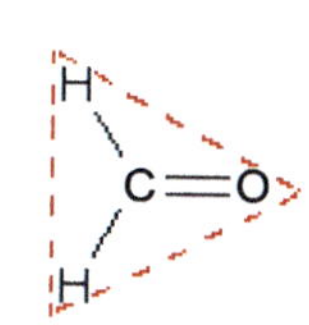
Linear config	02	180°	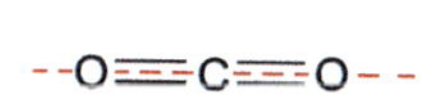

Source: Nicholls et al., Creative Commons License.

In the three cases, the center atom (carbon) lacks any valence electrons that are not involved in bonding. As a result, the configuration can be determined solely based on the number of atoms it is bonded to. When calculating the properties of water and ammonia molecules, it is necessary to consider the non-bonding electrons. For each scenario, there are four areas where electrons are concentrated in the outermost shell, resulting in an anticipated tetrahedral bond angle. Indeed, the arrangement of atoms, rather than electrons, determines the molecular structure. In this regard, ammonia is commonly described as pyramidal, rather than tetrahedral. Boron trifluoride, BF_3, lacks non-bonding electrons in valence and exhibits a trigonal atomic structure (Meyer & Richards, 1991).

Remember

Remember that molecular structure and bonding are intricately linked to the spatial arrangement of atoms within molecules, influencing their properties and reactivity.

Utilizing molecular models is the most effective approach to examining the three-dimensional structures of molecules. A wide variety of model kits are accessible to both students and chemists. Jmol, the model viewing applet, can provide an approximation of some of the beneficial characteristics of physical models. This potent visualization tool enables the user to manipulate a molecular structure in any desired manner (Rohrbaugh & Jurs, 1987).

Molecular dipole moments are a means by which the morphologies of molecules can be observed experimentally. A molecule that possesses a few polar covalent bonds can exhibit a dipole moment due to the accumulation of bond dipoles. Regarding water, it is established that the O-H covalent bond is polar because of the distinct electronegativities of oxygen and hydrogen. Due to the presence of two O-H bonds in water, the bond dipoles will come into contact and potentially create a measurable molecule dipole. The graphic illustrates four potential configurations of the O-H bonds (Nicholls et al., 2010) (Figure 3.1).

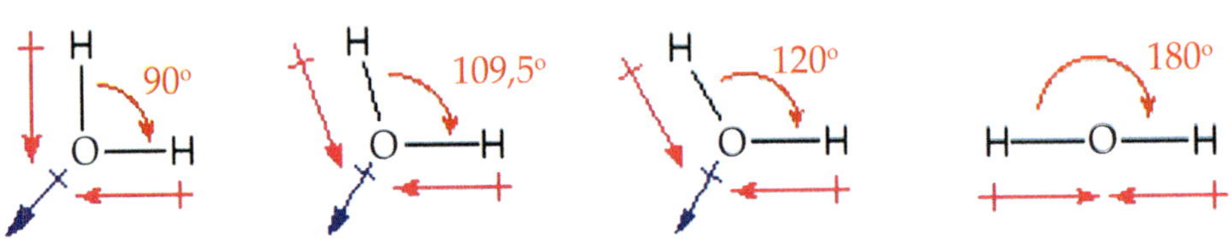

Figure 3.1. *Figure of different orientations of same elements.*

Source: homework, Creative Commons License.

Although the color magenta is used to depict the bond dipoles, blue is used to show the molecular dipole that is produced as a result of the bond dipoles. When the atoms are organized in a straight line (with a bond angle of 180°), the bond dipoles cancel each other out, which results in a molecular dipole that is equal to zero. It is possible that the amplitude of the molecular dipole would change for bond angles spanning from 120° to 90°, with the largest dipole occurring at a configuration representing 90°. It is possible to determine the configurations of methane (CH_4) and carbon dioxide (CO_2) by using the zero molecular dipole moments of both molecules comparably. As a result of the cancellation of bond dipoles, the molecular configurations of these molecules are required to be either tetrahedral or square-planar, and linear, respectively. Several different arguments have been used to verify the tetrahedral structure of methane, and the methane case provides useful insights into one of those justifications (Isihara, 1950).

The creation of a CH_3Cl molecule is the consequence of the substitution of one hydrogen atom with one atom of chlorine. Because hydrogen atoms in the tetrahedral, square-planar, and square-pyramidal configurations are structurally equivalent to one another, each of these configurations would result in the production of a single substitution product. On the other hand, in the trigonal-pyramidal configuration, one hydrogen atom, which is situated at the apex, demonstrates distinct structural properties in comparison to the other three hydrogen atoms, which are located at the base of the pyramid. In this scenario, the production of two separate CH_3Cl molecules would be the outcome of substitution if all the hydrogens met one another (Weisz, 1980). Concerning the process of di-substitution, if methane were to take on a tetrahedral form, the exclusive product that would be produced would be CH_2Cl_2. The alternative configurations, on the other hand, would result in the formation of two different compounds of CH_2Cl_2.

3.2. ISOMERS

Because a molecular formula does not, in most cases, provide a discrete representation of a single component, structural formulations for organic compounds are necessary. In the field of chemistry, the term "isomer" refers to compounds that have the same chemical formula but are distinct structures.

Constitutive isomers are compounds that contain molecules that are bonded together in fundamentally different ways. These compounds are referred to as constitutional isomers. The compound $C_4H_{10}O$ can be found in a total of seven different constitutional isomers. The formulas for the structural components of these isomers are presented in the table that follows. Among the $C_4H_{10}O$ compounds that have been recognized and those that have the potential to be discovered, these formulas display a structural property that is shared. These structures do not have any double or triple bonds, rings, or other types of bonds (Gordon, 1980) (Figure 3.2).

Structural Formulas for $C_4H_{10}O$ Isomers		
Kekule Formula	Condensed Formula	Shorthand Formula
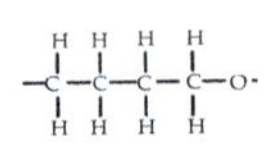	$CH_3(CH_2)_3OH$	
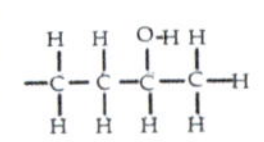	$C_2H_5CH(OH)CH_3$	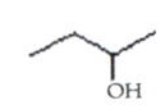
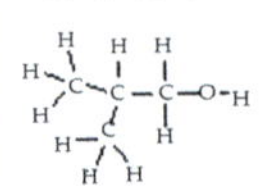	$(CH_3)_2CHCH_2OH$	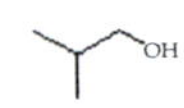
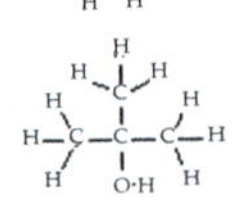	$(CH_3)_3COH$	
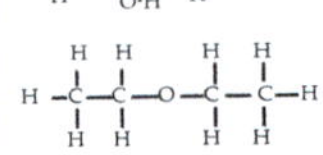	$(C_2H_5)_2O$	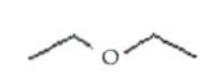
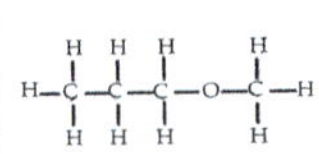	$CH_3(CH_2)_2OCH_3$	
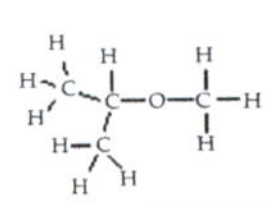	$(CH_3)_2CHOCH_3$	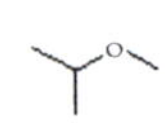

Figure 3.2. *Tabular representation of structural formulas of $C_4H_{10}O$ isomers.*

Source: Isihara, Creative Commons License.

There is a possibility of simplifying structural formulations without sacrificing the information that is being communicated. Subscript numbers are used to designate each distinct structural unit (group), including the hydrogens, in condensed structural formulas. These formulas do not include the bonds that are attached to each carbon, but they do indicate many substituents. The symbols for carbon and hydrogen are completely omitted from formulas that are written in shorthand. A chemical bond is represented by each linear segment, while carbon atoms are represented by the intersections and ends of the segments. To estimate the proper number of hydrogen atoms, the tetravalency of carbon is taken into consideration. When performing the computations, non-bonding valence shell electrons are not considered (Seburg et al., 1997). To acquire the ability to cognitively comprehend a three-dimensional arrangement based on two-dimensional mathematics, it is necessary to engage in consistent training and, in most cases, to make use of molecular models. Many kinds of model kits are available to both students and professional chemists on the market. It is strongly suggested that students who are just starting get one (Kelley et al., 2007).

Although the molecular formulae of constitutional isomers are identical, the physical and chemical properties of these isomers can vary greatly from one another (Rhuland et al., 1955).

- **Distinguishing Carbon Atoms:** During the process of analyzing structural formulas, it is frequently beneficial to discriminate between various groups of carbon atoms based on the structural characteristics that they possess. A carbon atom that is only bonded to one other carbon atom is referred to as a primary carbon, which has a degree of one. There are three different types of carbon atoms: tertiary (3°) and quaternary (4°) carbon atoms. Tertiary carbon atoms are connected to three other carbon atoms, while quaternary carbon atoms are connected to four other carbon atoms (Kier, 1986) (Figure 3.3).

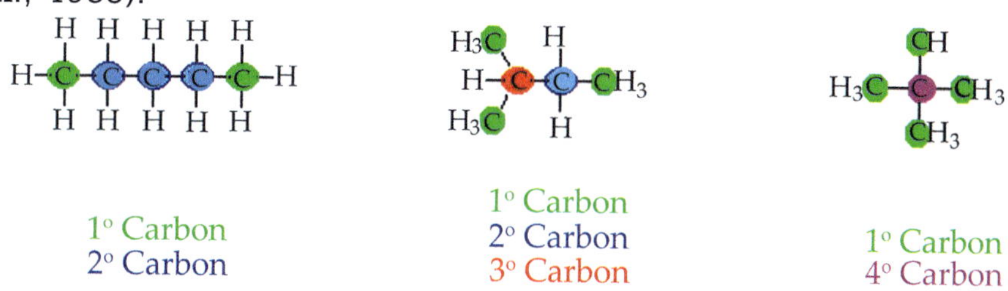

Figure 3.3. *Figure of these concepts are provided by the isomers.*

Source: Luman Learning, Creative Commons License.

These four groupings may have structural differences due to the molecular composition of their constituents. Due to the tetrahedral structure of tetravalent carbon, the formula on the right demonstrates that all four fundamental carbons are physically equal. Nevertheless, in the center formula, two primary carbons are comparable to the tertiary carbon on the left end, and there is a single primary carbon that is connected to the secondary carbon on the right end, which is a fundamentally distinct arrangement. Similarly, the formula on the left side of the equation comprises two 2°-carbons that are structurally equivalent and are situated next to the opposite ends of the chain. In addition, there is a 2°-carbon that is structurally different amid the chain between the double bonds (Newman & Hemmingson, 1995). It is possible to differentiate between atoms and groups that are structurally comparable and those that are not comparable by analyzing molecular symmetry. When it comes to obtaining knowledge in organic chemistry, one of the most basic aspects is the ability to recognize these kinds of structural distinctions. Over time, through constant practice and the accumulation of experience, mastery will be achieved (Esterhuysen & Frenking, 2011).

3.3. FORMULA ANALYSIS

While structural formulas are necessary for accurately describing organic compounds, it is intriguing to assess the information that can be derived solely from a molecular formula. Three beneficial guidelines can be enumerated.The maximum amount of hydrogen atoms that can form bonds with a specific number of carbon atoms is determined by the valence of carbon.

The greatest number of hydrogen atoms that can be linked to n carbons in hydrocarbons (compounds of carbon and hydrogen) is given by the formula 2n + 2, where n is an integer. For methane (CH_4), the value of n is 1, and 2n + 2 = 4. The derivation of this formula is apparent when considering a hydrocarbon composed of a linear arrangement of carbon atoms. In this case, the middle carbons will possess two hydrogen atoms each, whereas the two end carbons will have three hydrogen atoms each. Halogen atoms, which have a valence of 1, should be considered equal to hydrogen atoms (McNiel et al., 2010). This is demonstrated by the example of $C_3H_5Cl_3$, where the number of halogen atoms (n) is 3 and the total number of atoms (2n + 2) = 8, which is equal to the sum of the number of carbon atoms 5 and the number of hydrogen atoms 3. When nitrogen is present, each nitrogen atom (with a valence of 3) adds one to the maximum number of hydrogens.

Some Plausible Molecular Formulas	$C_7H_{16}O_3$, C_9H_{18}, $C_{15}H_{28}O_3$, $C_6H_{16}N_2$
Some Impossible Molecular Formulas	$C_8H_{20}O_6$, $C_{23}H_{50}$, $C_5H_{10}Cl_4$, $C_4H_{12}NO$

There is never a situation in which the total number of atoms in stable organic compounds that have an odd number of valence electrons is not an even number. Therefore, the number of unoccupied bonding locations must be an even number when atoms with an equal amount of valence electrons, such as carbon and oxygen, are bonded together in any configuration and quantity. This is because the valence electrons in these atoms are equal (Konishi et al., 2007). The total number of atoms will always be an even number if these sites are filled by monovalent atoms such as hydrogen, fluorine, chlorine, and other similar subatomic particles. The nitrogen atom has a valence of 3, which is an odd valence. The formation of a bond with carbon results in the creation of two more bonding sites, which guarantees that the total bonding pattern will continue to be even or odd (Schefe et al., 2006).

Some Plausible Molecular Formulas	$C_4H_4Cl_2$, C_5H_9OBr, $C_5H_{11}NO_2$, $C_{12}H_{18}N_2FCl$
Some Impossible Molecular Formulas	$C_5H_9O_2$, C_4H_5ClBr, $C_6H_{11}N_2O$, $C_{10}H_{18}NCl_2$

A direct correlation exists between the number of double-bonded molecules and rings that are contained in the structural formulas of stable compounds composed of carbon, hydrogen, and oxygen and the number of hydrogen atoms that are present in those compounds. Let's take a look at a hydrocarbon that has a chemical structure that is made up of a linear chain of four carbon atoms, namely $CH_3CH_2CH_2CH_3$. The chemical formula for this substance is C_4H_{10}, and it adheres to the 2n + 2 rule, which states that the largest number of hydrogens that are linked to a molecule is two. There is a requirement to eliminate two hydrogen atoms if the four carbon atoms are grouped in a circular configuration. To a similar extent, the removal of two hydrogen atoms is the consequence of the addition of a double bond, whilst the removal of four hydrogen atoms is the consequence of the addition of a triple bond (Gale et al., 2012) (Figure 3.4).

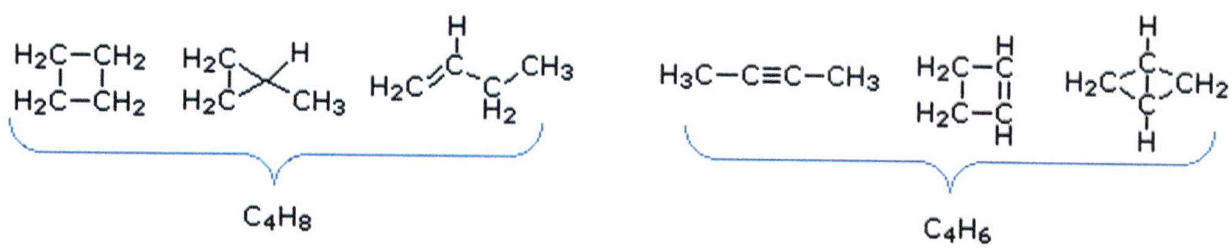

Figure 3.4. Illustration of molecular formula of butene and butadiene.

Source: Byjus, Creative Commons License.

An altered analysis is necessary when several nitrogen atoms or halogen substitutes are present. The equation can be expanded to include such compounds by using a few straightforward principles:

i. The existence of oxygen does not affect the bonding.

ii. All halogens in the molecular formula have to be substituted with hydrogen.

Every nitrogen in the equation needs to be substituted with a CH group (Borm et al., 2007).

3.4. RESONANCE

Kekulé structural formulae are crucial tools for comprehending organic chemistry. Nevertheless, several compounds and ions possess structures that cannot be adequately depicted by a solitary formula. For instance, sulfur dioxide (SO_2) and nitric acid (HNO_3) can each be represented by two equivalent formulations. To enhance clarity, the two bonds to oxygen that are open to interpretation are assigned distinct colors in these formulations (McDonnell et al., 2017).

1) Sulfur dioxide

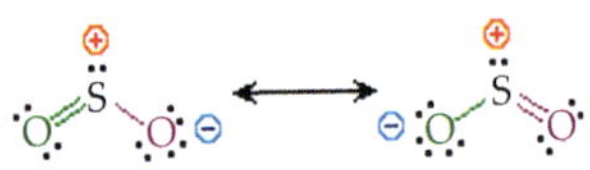

2) Nitric acid

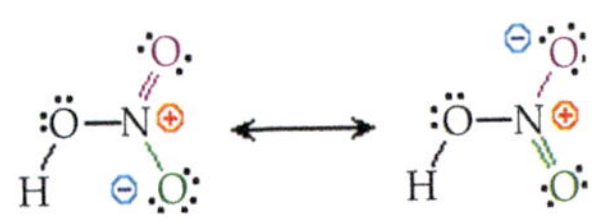

When compared to the single bond, the double bond between sulfur and oxygen would be shorter in length and stronger in strength. This would be the case if there were only one formula that was both precise and correct for sulfur dioxide. Based on experimental evidence, it has been demonstrated that this molecule exhibits a bond angle of 120° and possesses sulfur: oxygen links of equal length, with a measurement of 1.432 angstroms. As a result, a single formula is not adequate to represent its structure; rather, the true structure is a combination or average of the two formulas. A hybrid electronic structure is created by the process of resonance, which involves averaging the distribution of electrons over various hypothetical contributing structures, also known as canonical forms (Bleaney & Stevens, 1953).

Instances such as those described above are typical examples of resonance being used in severe situations. To create a stable compound, it is possible to formulate two electronic structures that are structurally and energetically equivalent to one another. Both structures, on the other hand, do not precisely or adequately represent the actual molecule. An enhancement in the stability of molecules can be achieved through resonance-induced electron delocalization in circumstances such as these (Van Noorden & Moelants, 1999).

3) Formaldehyde

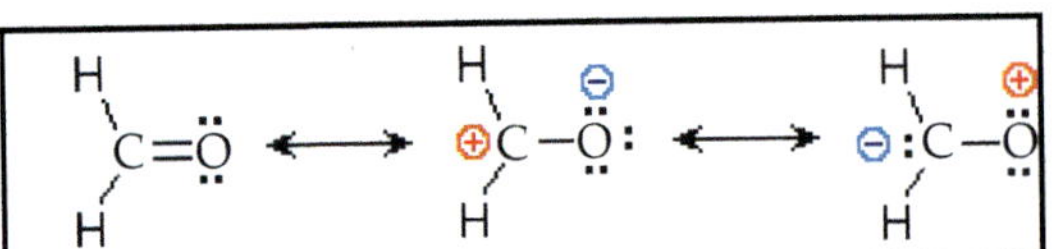

The electrical structures of most covalent compounds do not display the inadequacy that was discussed before. Consequently, methane (CH_4), water (H_2O), and acetylene (C_2H_2) are all capable of being adequately represented by formulas that fully satisfy the Kekulé equations. The notions of resonance, on the other hand, are extremely helpful in explaining the chemical behavior of many molecules of this kind. As a case study, the carbonyl group of formaldehyde, which is the double bond between carbon and oxygen, regularly goes through reactions that result in the production of additional products. A modest contribution from a dipolar resonance contributor can shed light on the course of these reactions, which displays the relationship between the two. This chemical unit is best represented by the first contributor on the left, which is the precise depiction. If the double bond is broken by the process of heterolysis, it results in the development of formal charge pairs, which are symbolized by the other two forms. An ideal charge distribution involves placing the positive charge on the element that has a lower electronegativity (carbon), and the negative charge on the atom that has a higher electronegativity (oxygen) (Bishop & Gladwell, 1963). Since this is the case, the central formula illustrates a structure that is more rational and long-lasting in comparison to the one that is shown on the right. To apply resonance to this scenario, it is necessary to compute a weighted average of these canonical structures. By far the most important contributor is the doubly bonded structure, followed by the middle structure, which is a secondary contributor, and the right-hand structure, which is believed to be an insignificant contributor. The fact that nucleophiles, which are electron donors, tend to join at this site can be explained by the presence of an electron-deficient carbon atom in the charge-separated contributor from the middle (Wikan, 1992).

Did you know?

Did you know that resonance in chemistry refers to the delocalization of electrons within molecules, often represented by resonance structures? This phenomenon stabilizes molecules and affects their chemical behavior.

To begin, let's take a moment to review the core concepts that underpin the resonance approach. An entire collection of Lewis and Kekulé structures is produced for a specific chemical, with the relative positions of all the constituent atoms being preserved throughout the process. All the canonical forms must have an equal number of paired and unpaired electrons. The following is a list of the canonical forms that need to be taken into consideration (Kubo & Tomita, 1954). To provide an accurate assessment of the influence that each of these canonical structures has on the real molecule, it is essential to consider the following aspects.

1. It refers to the number of covalent bonds that are present in a particular structure. There is a clear correlation between the significance and durability of the contributing structure and the strength of the bonding.
2. Separation of formal charges. (Unless additional influences are taken into consideration, the stability and relevance of the contributing structure are diminished as a result of charge separation).

3. Instability can occur when there is a high charge density present. Atoms that have a low electronegativity are the most suitable for accommodating positive charge, whereas atoms that have a high electronegativity are best for tolerating negative charge (Greenblatt, 1990).

For every single canonical contributor, the stability of a resonance hybrid is always superior to the stability of any individual contributor. Therefore, if a certain canonical form demonstrates considerably superior stability in comparison to all other forms, the hybrid that is produced will closely match that form in terms of the electrical and energetic properties that it possesses. It is an argument in support of the carbonyl group. When compared to both charge-separated structures, the C=O structure on the left hand demonstrates substantially stronger overall bonding. As a result, this structure is an extremely realistic representation of this functional group. The resonance hybrid, on the other hand, will experience extreme stability and exhibit different characteristics if two or more canonical forms share the same low-energy structures. These equations illustrate the existence of nitric acid and sulfur dioxide in the environment (Lauterbur, 1974).

4) Carbon monoxide		
5) Azide anion		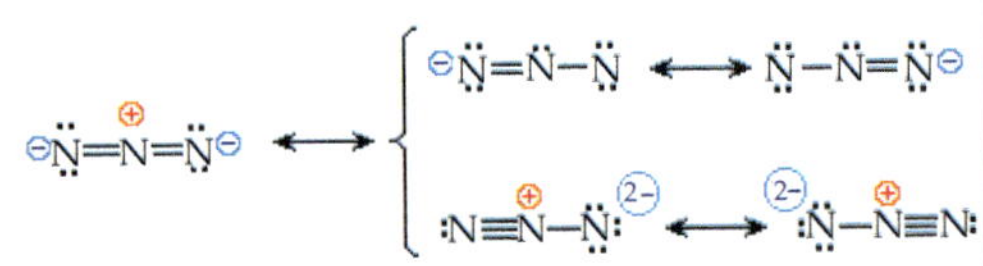

Both carbon monoxide and azide anion are examples of chemical substances that shall be investigated to illustrate these concepts. In every situation, the canonical form that is furthest to the left is the most stable. In the case of carbon monoxide, the separation of charges is not the primary focus; rather, the emphasis is focused on the additional bonding. Additionally, the structure that is double-bonded has a carbon atom that has a shortfall of electrons in its valence shell, precisely having six electrons. There is a similar destabilizing characteristic shared by the two azide canonical forms that are located on the top row of the bracket. This characteristic is the difference in the number of bonds, which is three bonds in the leftmost structure and four bonds in the other structure (Alston et al., 1960).

Every single example that is presented on this page exemplifies a significant limitation that must be taken into consideration whenever resonance is utilized. Atoms do not shift relative to one another inside the context of the common structural structure. In this process, the only particles that are displaced are electrons (Close & Page, 2004).

3.5. ORBITALS

3.5.1. Atomic and Molecular Orbitals

To develop a more comprehensive understanding of covalent bonding, it is necessary to consider the valence shell atomic orbitals. The second-period elements, such as carbon, nitrogen, and oxygen, have been assigned the orbitals 2s, $2p_x$, $2p_y$, and $2p_z$ (Figure 3.5).

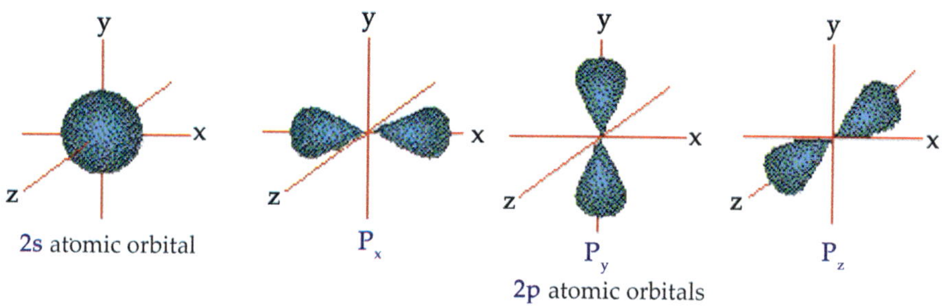

Figure 3.5. *Example of atomic and molecular orbitals.*

Source: Tutors Globe, Creative Commons License.

The valence shell of carbon has the following electron configuration: $2s^2$, $2p_x^{\ 1}$, $2p_y^{\ 1}$, and $2p_z^{\ 0}$. In the context of covalent bonding, carbon would only be able to form two bonds with this structure, which shows that its capacity for bond formation is restricted (Vozzi et al., 2011).

The outermost electron shell would have a total of six electrons, which is two electrons less than a complete set of eight electrons. Despite this, the tetrahedral configurations of methane and carbon tetrachloride demonstrate that carbon can form four bonds that are indistinguishable from one another, which ultimately results in the production of the necessary octet.

An orbital hybridization model was presented by Linus Pauling to shed light on the phenomenon of covalent bonding. This model involves the reorganization of the electrons in the valence shell of carbon (Edmiston & Ruedenberg, 1965).

3.5.2. Hybrid Orbital

To elucidate the arrangement of methane (CH_4), the 2s and three 2p orbitals transform four indistinguishable hybrid atomic orbitals. These hybrid orbitals possess a composition of 25% s character and 75% p character and are denoted as sp^3. The hybrid orbitals possess a distinct orientation, with all four naturally aligned in a tetrahedral manner (Foster & Weinhold, 1980).

3.5.3. Molecular Orbitals

On the other hand, the shared electron pairs of covalently bonded atoms may be considered to occupy molecular orbitals (MO), whereas the valence electrons of atoms are present in atomic orbitals (AO). This can be done individually or in combination. It is common practice for the combining of n atomic orbitals to result in the production of n molecular orbitals by default. To provide a plain explanation of the production of molecular orbitals (MO), the hydrogen molecule is present. The diagram depicts the process of combining two 1s atomic orbitals, which ultimately leads to the production of a molecular orbital with a low-energy sigma (⊠) bonding and an antibonding orbital with a greater energy. Because of the presence of a pair of electrons that have opposite spins, the molecular orbital (MO) is filled with electrons, which leads to the creation of a covalent bond (Coulson & Rushbrooke, 1940).

Tip

Hybrid orbitals arise from the mixing of atomic orbitals to form new orbitals suitable for bonding. Understanding hybridization helps predict molecular shapes and bond angles in complex molecules.

The nomenclature that is used for molecular orbitals is quite like the notation that is used for atomic orbitals. Because of this, s-orbitals are characterized by a spherical form that is symmetrical around a single nucleus, whereas ⊠-orbitals are characterized by a cylindrical shape that is symmetrical and surrounds two or more nuclei. When atoms of the second period form bonds with one another, they construct molecular orbitals by employing p-orbitals or hybrid atomic orbitals that have features like those of p-orbitals. One example is the sigma molecular orbital, which is responsible for bonding two fluorine atoms (Lennard, 1949). This orbital is generated by the overlapping of p-orbitals, as described in part A in Figure 3.6.

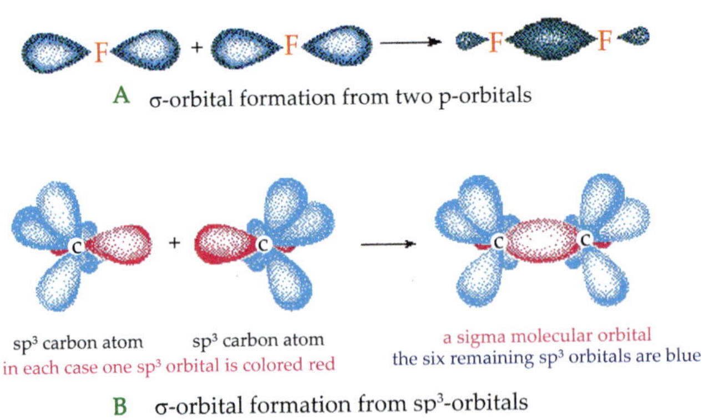

Figure 3.6. *Illustration of sp^3 orbitals.*

Source: Coulson, Creative Commons License.

Another variant of molecular orbital (referred to as the π orbital) can be created by the lateral overlap of two p-orbitals, as illustrated in portion A of the Figure 3.7.

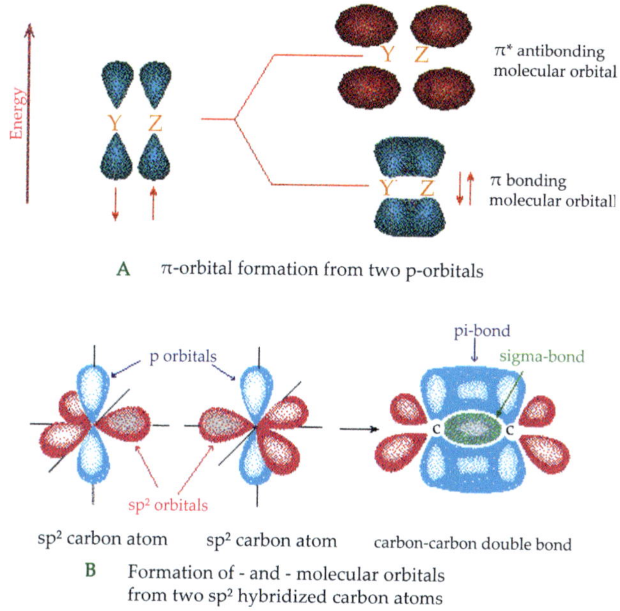

Figure 3.7. *Figure of lateral overlap of two p-orbitals.*

Source: Geeks for Geeks, Creative Commons License.

Molecular orbitals are created through a process that is indeed more complex than the isolated instances that were presented earlier. This process involves the combination of atomic orbitals. The building of an orbital correlation diagram is a requirement of modern molecular orbital theory, even though these models help clarify the composition and responsiveness of a wide variety of organic molecules. Two diagrams depicting the simple diatomic elements F_2 and N_2 will be displayed above. Because of the limited orbital overlap, the atomic orbitals of 1s and 2s do not contribute to total bonding. This leads to the cancellation of sigma bonding and antibonding components, which are both beneficial to the bonding process. In each of these instances, the combination of three 2p atomic orbitals results in the formation of a sigma and two pi-molecular orbitals. Each pair of orbital pairs functions as a bonding and antibonding pair. The occupancy of antibonding orbitals is what determines the order in which bonds should be formed. The tiny alteration in the energy of the ⊠2p bonding orbital, in comparison to the two corresponding π-bonding orbitals, is a result of s-p hybridization, which is not relevant to the current subject (Boys, 1960).

Oxygen (O_2) is an example that shows the advantages that the molecular orbital technique brings to the process of bonding. The paramagnetic properties of this straightforward diatomic molecule are depicted precisely by the correlation diagram. Similarly, the orbital correlation diagram for methane highlights the disparity in electron density that is projected by molecular orbital calculations in comparison to that of the localized bond model. In this model, the p-orbitals are shown as red and blue spheres or ellipses. These spheres or ellipses indicate unique phases that are specified by the mathematical wave equations for these orbitals (Pascual et al., 2000).

Only two hybrid orbitals are necessary for the formation of sigma bonds in carbon atoms that have only two partners with which they can form bonds. These hybrid orbitals, which are referred to as sp orbitals, are orientated in opposite directions at an angle of 180° from each other respectively. Following the sp-hybridization of every atom, there remain two p-orbitals that are not utilized due to their absence. A triple bond is produced because of the overlapping of these p-orbitals, which results in the formation of two pi-bonds in addition to the sigma bond (Inadomi et al., 2002).

3.6. MOLECULAR GEOMETRY

The Lewis electron-dot structures do not have any geometrical value other than illustrating the arrangement of atoms and their connections. However, by slightly expanding the basic concept of shared-electron pairs, one may effectively explain and forecast the arrangement of bonds surrounding a certain atom in a diverse range of scenarios (Bader et al., 1988).

3.6.1. Electron-Pair Repulsion

The VSEPR model discussed here specifically examines the electron pairs, both bonding and nonbonding, located in the outermost shell of an atom that form bonds with two or more additional atoms. Like all electrons, these particles exist in certain areas of space that can be conceptualized as electron clouds. These clouds represent regions of negative electric charge, sometimes referred to as orbitals. The exact nature of these orbitals can be better understood through more comprehensive theories (Gillespie, 1967). When it comes to chemical bonding, the covalent model assumes that the electron pairs that are responsible for bonding are concentrated in the space that exists between the atoms that are connected. The fundamental idea behind the VSEPR hypothesis is that regions with a negative electric charge would be attracted to one another and display repulsion towards one another. This would cause these regions to tend to keep as much distance as possible from one another, including the chemical bonds that they form. As a result, the two-electron clouds that are present in a basic triatomic molecule, AX_2, will extend out in opposite directions. When two bonding orbitals are angularly separated by 180°, they are positioned in such a way that they are removed from each other to the greatest extent feasible. Because of this, the two chemical bonds will stretch in different directions, which will ultimately result in a linear molecule (Burdett & Fassler, 1991). These additional regions of negative charge will behave in a manner that is comparable to that of the bound atoms if the center atom additionally possesses one or more pairs of electrons that are not bonded to one another. In the outermost electron shell, the orbitals that house the various bonding and nonbonding pairs will extend outward from the central atom in orientations that minimize the mutual repulsions that exist between them. The term "expanded octet" is widely used to describe the situation in which the center atom has partially full d-orbitals. This allows the center atom to potentially store five or six electron pairs without being filled (Andres et al., 2016).

3.6.2. Diagonal and Trigonal Coordination

3.6.2.1. Linear Molecules

According to what was stated earlier, a triatomic molecule of the AX_2 type possesses two bonding orbitals that are separated by 180°, which results in the molecule having a linear shape. $BeCl_2$ and CO_2 are two instances of triatomic molecules that exhibit a linear shape, as predicted by the VSEPR theory. However, due to the existence of an inadequate number of electrons, $BeCl_2$ does not satisfy the octet rule (Rost et al., 1992). Carbon dioxide is another example. When the electron dot formula for carbon dioxide is written out, it becomes immediately apparent that the C-O bonds present in the compound are double bonds. This does not have any impact on the VSEPR hypothesis; the core carbon atom continues to be connected to two other atoms, and the electron clouds that connect the two oxygen atoms continue to be positioned 180° apart (Brown et al., 1975).

3.6.2.2. Trigonal Molecules

In a molecule like BF_3, which belongs to the AX_3 molecular geometry, three zones of electron density spread outward from the core atom (Lauterbur & Ramirez, 1968). The repulsion between these objects will be minimized when the angle between either of them is 120°, (360° / 3 = 120°).

3.6.3. Tetrahedral Coordination

Methane, CH_4, consists of a single carbon atom that is chemically linked to four hydrogen atoms. What bond angle would result in the maximum possible distance between the electron clouds linked to these bonds? Following the same pattern as the previous two situations, where the bond angles were 360°/2=180° and 360°/3=120°, one may assume that the bond angle for 360°/4 would be 90°. However, this assumption would be incorrect. The latter calculation would be accurate if all the atoms were obliged to be on the same plane, but in this case, there is no such limitation (Curie et al., 1974).

The tetrahedral coordination of carbon in most organic compounds was discovered in the nineteenth century using geometric concepts and chemical evidence. This discovery came before the invention of direct methods for detecting the shapes of molecules. The chemical formula CH_2Cl_2 can be used to describe dichloromethane, which is the solitary instance of the compound (Blanchard et al., 2005).

If the coordination around the carbon atom were square, it would consequently be necessary for there to be two isomers of CH_2Cl_2, as illustrated by the pair of structures that were presented earlier. The difference in distances between the two chlorine atoms leads to the formation of various physical properties, which in turn makes it possible to differentiate and separate the two isomers and distinguish them from one another (Roe & Pang, 1999).

3.6.3.1. Tetrahedrally-Coordinated Carbon Chains

Alkane chains get the general formula $H_3C-(CH_2)n-CH_3\ (orCnH2n+2)$ are possible to be built up; an instance of pentane, C_5H_12, is shown in Figure 3.8.

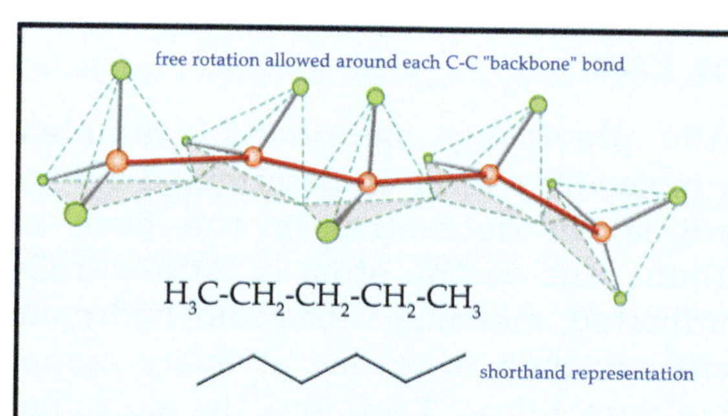

Figure 3.8. *Image of alkane chains.*

Source: Libre texts, Creative Commons License.

It is important to observe that these hydrocarbons, commonly referred to as "straight chain hydrocarbons," really contain a carbon backbone structure that is not truly straight. This is depicted by the zig-zag image commonly used to represent hydrocarbon structures (Xu et al., 2011).

3.6.3.2. Coordination Geometry and Molecular Geometry

The coordination number is the count of electron pairs that surround the central atom. This number is independent of the actual position of the atom inside the molecule. If all the electron pairs that surround the core atom are dispersed among the atoms that are adjacent to it, then the coordination geometry and the molecular geometry will be identical.

Molecular geometry is less complicated than coordination geometry if there is at least one electron pair around the central atom that is not shared with any other atom in the vicinity. This type of pair is referred to as a lone pair. It is possible to analyze a diagram of the coordination geometry to ascertain the molecular geometry (Hosseinzadeh & Ahmadi, 2022).

3.6.3.3. Tetrahedral Coordination with Lone Pairs

In the previously mentioned examples, the molecule's form is determined by its coordination geometry. In the case of methane, the carbon atom is tetrahedrally coordinated, meaning it has four hydrogen atoms attached to it, one at every corner of the tetrahedron. Therefore, the molecular shape of methane is also tetrahedral (Millet et al., 2001).

Bonding patterns are commonly represented using "generic" formulas like AX_4, AX_2E_2, etc., where "X" represents bonding pairs and "E" represents lone pairs. This convention is commonly referred to as the "AXE Method."

If the valence shell of the central atom comprises nonbonding electrons, the bonding geometry is unlikely to be tetrahedral. The presence of nonbonding electrons in orbitals results in spatial occupation and repulsion of other orbitals. When determining the coordination number of the central atom, it is necessary to consider both the atoms that are bonded to it and the pairs of electrons that are not involved in bonding (Krivovichev, 2004).

Two of the coordination locations are filled by the electron pairs that make up the O–H bonds, while the other two are taken by the non-bonding pairs. Because the oxygen atom is coordinated tetrahedrally, the bonding geometry of the H_2O molecule is defined as twisted (Lawton et al., 1974).

Bonding and non-bonding electron orbitals exhibit a significant distinction. Due to the absence of an atomic nucleus at its far end, a nonbonding orbital lacks the force to attract the electron cloud towards it. As a result, the charge in this orbit will be more concentrated in proximity to the core atom (Chidambaram et al., 1964).

This image was generated by a computerized simulation utilizing the comprehensive molecular orbital model, which will be discussed in detail in the upcoming session. Despite being electrically neutral, the water molecule exhibits non-uniform electrical distribution due to the presence of non-bonding electrons. This results in a greater percentage of negative charge (blue hue) at the oxygen end, causing the hydrogen side to be comparatively positive (red). The presence of this charge imbalance is responsible for many of the unusual features exhibited by water (Rieger & Mudring, 2007).

3.6.3.4. Ammonia: AX_3E

In the electron-dot structure of NH_3, a single pair of nonbonding electrons is in the valence shell of the atom of nitrogen. This indicates that there are three atoms chemically connected to the nitrogen atom, along with one unshared pair of electrons. As a result, the coordination number of nitrogen is four, which is the same as the coordination number in H_2O. Consequently, one can anticipate that each of the three hydrogen atoms will be positioned at the vertices of a tetrahedron, with the nitrogen atom at its center. The lone pair orbital of the NH_3 molecule will be directed toward the fourth corner of the tetrahedron (Mashruk et al., 2023). However, since that position will be unoccupied, the NH_3 molecule cannot have a tetrahedral shape. On the contrary, it takes on a pyramidal form. To be more accurate, the shape can be described as a trigonal pyramid, which means it has a triangle base. All hydrogen atoms are coplanar, but the nitrogen atom is positioned either above, below, or to the side. It is important to note that molecules cannot perceive spatial orientation using terms like "above" or "below." The larger orbital that contains the non-bonding electrons exerts pressure on the bonding orbitals, causing them to come closer together, resulting in H–N–H bond angles of approximately 107° (Andersson et al., 2023).

3.6.3.5. Central Atoms with Five Bonds

Compounds of the AX-5 type can be formed from elements belonging to Group 15. Among the chemicals that fall within this category are PCl-5 and AsF-5. Assuming that there are five electron pairs, what are the potential spatial configurations that would minimize the repulsions between them? When it comes to coordination numbers 2, 3, 4, and 6, one can imagine the electron pairs arranging themselves over the surface of a sphere in the most extensive feasible manner. In the case of bigger numbers, the geometries that are produced are equivalent to regular polyhedra that have the same number of sides. The lack of a regular polyhedron that has five vertices is the problem with coordination number 5, which is the fifth coordinate (Farnham et al., 1988).

3.6.3.6. Regular Polyhedra

In 1758, the eminent mathematician Euler demonstrated that there were only five regulars convex polyhedra, which are commonly referred to as platonic solids. These include the tetrahedron, which has four triangular faces, the octahedron with six triangular faces, the icosahedron with 20 triangular faces, the cube with six square faces, and the dodecahedron with 12 pentagonal faces. Chemical examples of all types are well-documented. One notable example is the first icosahedral molecule, LaC_{60}, which was synthesized in 1986. In this molecule, the lanthanum (La) atom is surrounded by 20 carbon (C) atoms in its immediate vicinity (MÁTHÉ & BÖRÖCZKY, 1999).

The molecular geometry of PCl_5 and molecules with a similar structure is classified as a trigonal bipyramid (Figure 3.9). This is comprised of two pyramids with triangular bases that are connected by their bases. Three chlorine atoms are positioned in the same plane as the center phosphorus atom, known as the equatorial position. The remaining two chlorine atoms are located above and below this plane, referred to as the axial positions. Equatorial and axial atoms exhibit distinct geometric interactions with their neighboring atoms, resulting in small variations in their chemical behavior (Deza & Grishukhin, 2001).

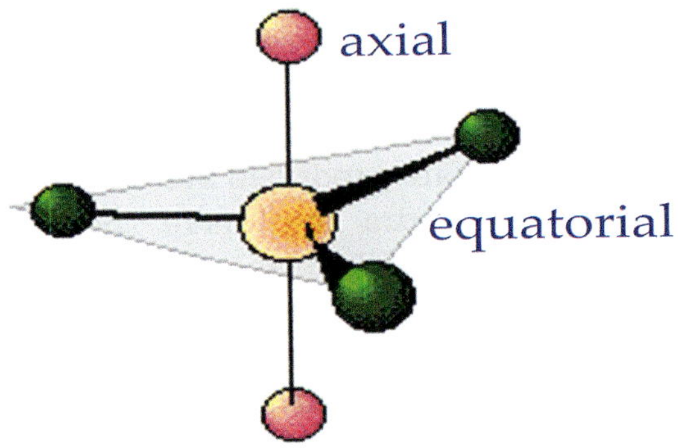

Figure 3.9. *Image of trigonal bipyramid structure.*

Source: Chem 1, Creative Commons License.

In molecules with 5-coordinated structures and lone pairs, the non-bonding orbitals, which are located closer to the central atom and therefore more susceptible to repulsion from other orbitals, will predominantly occupy the equatorial plane. This will position them at 90° angles relative to a maximum of two bonding orbitals that are aligned along the axis (Brown, 1990).

Based on this logic, one can anticipate that a molecule with the formula AX_4E (where the central atom A is connected to four other atoms "X" and one nonbonding electron pair), like SF_4, will have a "see-saw" structure. Adding more nonbonding pairs instead of bonded atoms simplifies the triangular bipyramid coordination, resulting in even simpler molecular shapes, as demonstrated in Figure 3.10 (Cunningham & Pellicer, 2016).

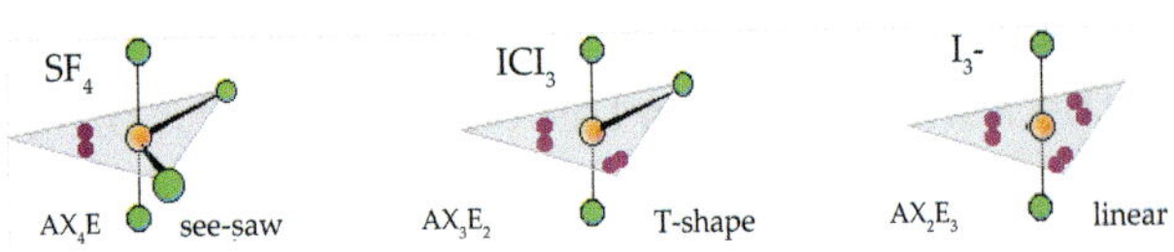

Figure 3.10. *Illustration of the triangular bipyramid coordination bonded atoms.*

Source: Study, Creative Commons License.

3.6.4. Octahedral Coordination

The arrangement of six electron pairs in an octahedron is such that they are steered toward the corners, minimizing repulsion. An octahedron is not as intricate a shape as the name might suggest; it is merely the combination of two square-based pyramids connected at their bases (Murray et al., 1972) (Figure 3.11).

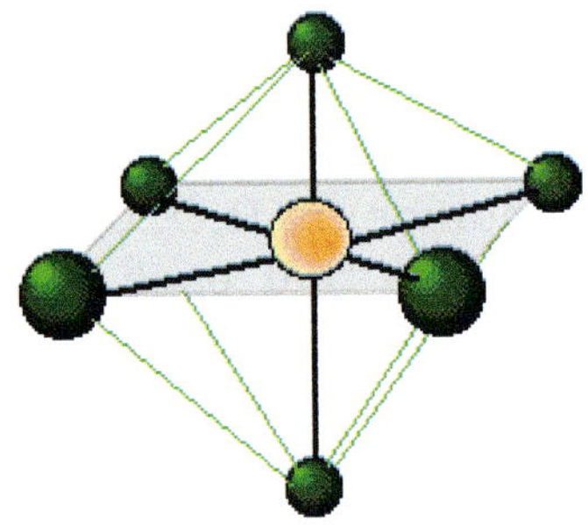

Figure 3.11. *Diagram of octahedrally-coordinated molecule.*

Source: Geeks for Geeks, Creative Commons License.

Among the three equivalent planes indicated by a four-fold symmetry axis, the shaded plane seen in this octahedrally-coordinated molecule is just one of them. The ligands in an AX_6 molecule are geometrically equal, meaning there are no distinct axial and equatorial locations.

Initially, one could perceive a coordination number of six as quite uncommon. It unequivocally contradicts the octet rule, and just a handful of molecules (such as SF_6) have a hexavalent core atom. Interestingly, this coordination number is frequently seen in the field of inorganic chemistry. There are two primary factors contributing to this:

1. Numerous transition metal ions establish coordinate covalent connections with atoms that donate lone-pair electrons, such as nitrogen (in ammonia) and oxygen

(in water). Transition elements can contain up to six electron pairs surrounding the core atom due to their outer arrangement of d10s^2. A coordination number of 6 is frequently observed in transition metal hydrates, such as $Fe(H_2O)^{63+}$.

2. While the center atom in most compounds is often connected to fewer than six other atoms, there are often enough lone-pair electrons to make the overall number of electron pairs equal to six (Schlemper & Hamilton, 1966).

3.6.4.1. Octahedral Coordination with Lone Pairs

There are significant examples of central atoms that have six coordination sites and possess one, two, or three electron pairs that are not shared by any other entities. Therefore, the octahedral coordination observed around the core atom is present in all three of the compounds that are presented below. It is essential to point out that the orientation of the shaded planes that can be observed in the two photographs that are located to the right is completely arbitrary. As a result of the fact that each of the six vertices of an octahedron is identical, it is possible to draw the planes in any of the three possible vertical orientations (Gao et al., 2021).

3.6.5. VSEPR Theory

Given its extraordinary simplicity, the VSEPR model possesses a surprising capacity for effectiveness. Molecular structures can be predicted using this method, and the following is an explanation of how it can be applied:

1. The electron pairs that surround a central atom exhibit repulsion towards one other; this repulsion is minimized when the orbitals housing these electron pairs are positioned in a manner that maximizes their distance from each other (Habiddin et al., 2023).
2. A polyhedron that has the same number of vertices as the number of electron pairs that surround it (also known as the coordination number) is the one that determines the coordination geometry that surrounds the central atom. One of the regular polyhedra will always be the shape, except the number 5, as well as the straightforward circumstances of two and three (Nawawi, 2021).
3. The molecular shape will be less complicated than the coordination polyhedron if there are electron pairs that do not form bonds between the atoms.
4. The concentration of orbitals that are close to the core atom is higher for those orbitals that have electrons that do not form bonds with other electrons. Furthermore, in comparison to bonding pairs, they have a larger ability to repel other orbitals (Malcolm & Popelier, 2003).

VSEPR theory is unable to provide accurate details, even though it is frequently reliable in forecasting the general shapes of most molecules. The bond angle in H_2O is 104.5°, but in H_2S it is roughly 90°. For example, it does not explain why this is the case. Given that the focus is on electrical repulsions, where the precise properties of the orbitals that hold the electrons and, consequently, the bonds themselves are not taken into consideration, this is not surprising at all.

ACTIVITY 3.1.

Apply VSEPR (valence shell electron pair repulsion) theory to predict the molecular geometry of various molecules. Use models or diagrams to visualize electron pair repulsion and determine molecular shapes based on the number of bonding and non-bonding electron pairs.

SUMMARY

- Chapter 3 focuses on the molecular structure and bonding of atoms, providing an in-depth look at how atoms combine to form molecules and the factors that influence their shapes and stability. It begins by exploring molecular shapes, including the various types of isomers, such as structural and stereoisomers, and the methods used for formula analysis. The concept of resonance is introduced to explain the delocalization of electrons within molecules, which contributes to their stability.
- The chapter further examines atomic and molecular orbitals, detailing the formation of hybrid orbitals and their significance in bonding. Molecular geometry is discussed with a focus on electron-pair repulsion, which influences the arrangement of atoms in a molecule. Different coordination geometries, such as diagonal, trigonal, tetrahedral, and octahedral, are explained along with the Valence Shell Electron Pair Repulsion (VSEPR) theory. This theory helps predict the shapes of molecules based on the repulsion between electron pairs.

REVIEW QUESTIONS

1. What are the primary factors that determine the shape of a molecule?
2. How do structural, geometric, and optical isomers differ from each other?
3. Explain the importance of resonance in the stability of molecular structures.
4. Describe the difference between atomic orbitals and molecular orbitals.
5. How does hybridization of orbitals affect the geometry of a molecule?
6. What is the VSEPR theory and how does it help predict molecular shapes?
7. Discuss the significance of electron-pair repulsion in determining molecular geometry.

MULTIPLE CHOICE QUESTIONS

1. **What is the main concept behind VSEPR theory?**
   a. Minimization of electron-electron repulsion
   b. Maximization of molecular mass
   c. Equal distribution of atomic masses
   d. Preservation of atomic integrity
2. **Which type of isomerism involves different spatial arrangements around a double bond?**
   a. Structural isomerism
   b. Geometric isomerism
   c. Optical isomerism
   d. Resonance isomerism
3. **What is the significance of resonance in molecular structures?**
   a. It increases the molecular weight
   b. It provides multiple stable structures for a molecule
   c. It changes the atomic number
   d. It enhances the physical state of a molecule
4. **Which orbitals are involved in hybridization to form sp3 hybrid orbitals?**
   a. One s and one p orbital
   b. One s and two p orbitals
   c. One s and three p orbitals
   d. Two s and two p orbitals
5. **In the context of molecular orbitals, what does the term "bonding orbital" refer to:**
   a. An orbital that decreases the energy of the molecule when occupied
   b. An orbital that increases the energy of the molecule when occupied
   c. An orbital with no impact on molecular energy
   d. An orbital that defines the magnetic properties of a molecule
6. **Which type of molecular geometry is associated with tetrahedral coordination?**
   a. Linear
   b. Trigonal planar
   c. Tetrahedral
   d. Octahedral
7. **What effect does electron-pair repulsion have on the shape of a molecule?**
   a. It compresses the angles between bonds
   b. It forces the molecule into a flat shape

c. It maximizes the distance between electron pairs

d. It eliminates lone pairs from the molecule

Answers to Multiple Questions

1. (a); 2. (b); 3. (b); 4. (c); 5. (a); 6. (c); 7. (c).

REFERENCES

1. Alston, M., Alvarez, L. W., Eberhard, P., Good, M. L., Graziano, W., Ticho, H. K., & Wojcicki, S. G. (1960). Resonance in the Λ π system. *Physical Review Letters, 5*(11), 520.

2. Andersson, K., Delin, S., Pedersen, J., Hafner, S. D., & Nyord, T. (2023). Ammonia emissions from untreated, separated and digested cattle slurry–Effects of slurry type and application strategy on a Swedish clay soil. *Biosystems Engineering, 226*, 194–208.

3. Andres, J., Berski, S., & Silvi, B. (2016). Curly arrows meet electron density transfers in chemical reaction mechanisms: From electron localization function (ELF) analysis to valence-shell electron-pair repulsion (VSEPR) inspired interpretation. *Chemical Communications, 52*(53), 8183–8195.

4. Auf der Heyde, T. P., & Buergi, H. B. (1989). Molecular geometry of d8 five-coordination. 1. Data search, description of conformation, and preliminary statistics. *Inorganic Chemistry, 28*(21), 3960–3969.

5. Bader, R. F., Gillespie, R. J., & MacDougall, P. J. (1988). A physical basis for the VSEPR model of molecular geometry. *Journal of the American Chemical Society, 110*(22), 7329–7336.

6. Birk, J. P., & Kurtz, M. J. (1999). Effect of experience on retention and elimination of misconceptions about molecular structure and bonding. *Journal of Chemical Education, 76*(1), 124.

7. Bishop, R. E. D., & Gladwell, G. M. L. (1963). An investigation into the theory of resonance testing. *Philosophical Transactions of the Royal Society of London. Series A, Mathematical and Physical Sciences, 255*(1055), 241–280.

8. Blanchard, S., Neese, F., Bothe, E., Bill, E., Weyhermüller, T., & Wieghardt, K. (2005). Square planar vs tetrahedral coordination in diamagnetic complexes of nickel (II) containing two bidentate π-radical monoanions. *Inorganic Chemistry, 44*(10), 3636–3656.

9. Bleaney, B., & Stevens, K. W. H. (1953). Paramagnetic resonance. *Reports on Progress in Physics, 16*(1), 108.

10. Borm, G. F., Fransen, J., & Lemmens, W. A. (2007). A simple sample size formula for analysis of covariance in randomized clinical trials. *Journal of Clinical Epidemiology, 60*(12), 1234–1238.

11. Boys, S. F. (1960). Construction of some molecular orbitals to be approximately invariant for changes from one molecule to another. *Reviews of Modern Physics, 32*(2), 296.

12. Brown, C. S. (1990). Capacity of the regular polyhedra. *Computers & Mathematics with Applications, 20*(1), 43–56.

13. Brown, J. M., Hougen, J. T., Huber, K. P., Johns, J. W. C., Kopp, I., Lefebvre-Brion, H., & Zare, R. N. (1975). The labeling of parity doublet levels in linear molecules. *Journal of Molecular Spectroscopy, 55*(1–3), 500–503.

14. Burdett, J. K., & Fassler, T. F. (1991). Pair potentials, electron-pair repulsions, and the geometries of molecules. *Inorganic Chemistry, 30*(14), 2859–2868.

15. Chiang, C. L., Xu, C., Han, Z., & Ho, W. (2014). Real-space imaging of molecular structure and chemical bonding by single-molecule inelastic tunneling probe. *Science, 344*(6186), 885–888.

16. Chidambaram, R., Sequeira, A. T., & Sikka, S. K. (1964). Neutron-diffraction study of the structure of potassium oxalate monohydrate: Lone-pair coordination of the hydrogen-bonded water molecule in crystals. *The Journal of Chemical Physics, 41*(11), 3616–3622.

17. Close, F. E., & Page, P. R. (2004). The D∗ 0D̄0 threshold resonance. *Physics Letters B, 578*(1–2), 119–123.

18. Coulson, C. A., & Rushbrooke, G. S. (1940). Note on the method of molecular orbitals. *Mathematical Proceedings of the Cambridge Philosophical Society, 36*(2), 193–200.

19. Cunningham, G., & Pellicer, D. (2016). Classification of tight regular polyhedra. *Journal of Algebraic Combinatorics, 43*, 665–691.

20. Curie, D., Barthou, C., & Canny, B. (1974). Covalent bonding of Mn^{2+} ions in octahedral and tetrahedral coordination. *The Journal of Chemical Physics, 61*(8), 3048–3062.

21. Deza, M., & Grishukhin, V. (2001). Face-regular bifaced polyhedra. *Journal of Statistical Planning and Inference, 95*(1–2), 175–195.

22. Edmiston, C., & Ruedenberg, K. (1965). Localized atomic and molecular orbitals. II. *The Journal of Chemical Physics, 43*(10), S97–S116.

23. Esterhuysen, C., & Frenking, G. (2011). Distinguishing carbones from allenes by complexation to AuCl. *Chemistry–A European Journal, 17*(36), 9944–9956.

24. Farnham, W. B., Dixon, D. A., & Calabrese, J. C. (1988). Novel fluorine-bridged polyfluorinated iodine structures. The presence of fluorine as the central atom in a five-center, six-electron bond. *Journal of the American Chemical Society, 110*(25), 8453–8461.

25. Foster, J. P., & Weinhold, F. (1980). Natural hybrid orbitals. *Journal of the American Chemical Society, 102*(24), 7211–7218.

26. Gale, C., Logan, K. M., Santhakumaran, S., Parkinson, J. R., Hyde, M. J., & Modi, N. (2012). Effect of breastfeeding compared with formula feeding on infant body composition: A systematic review and meta-analysis. *The American Journal of Clinical Nutrition, 95*(3), 656–669.

27. Gao, L., Yadgarov, L., Sharma, R., Korobko, R., McCall, K. M., Fabini, D. H., & Yaffe, O. (2021). Metal cation s lone-pairs increase octahedral tilting instabilities in halide perovskites. *Materials Advances, 2*(14), 4610–4616.

28. Gillespie, R. J. (1967). Electron-pair repulsions and molecular shape. *Angewandte Chemie International Edition in English, 6*(10), 819–830.

29. Gordon, M. S. (1980). The isomers of silacyclopropane. *Chemical Physics Letters, 76*(1), 163–168.

30. Grant, J. A., Gallardo, M. A., & Pickup, B. T. (1996). A fast method of molecular shape comparison: A simple application of a Gaussian description of molecular shape. *Journal of Computational Chemistry, 17*(14), 1653–1666.

31. Greenblatt, S. (1990). Resonance and wonder. *Bulletin of the American Academy of Arts and Sciences, 43*(4), 11–34.

32. Haaland, A. (1979). Molecular structure and bonding in the 3d metallocenes. *Accounts of Chemical Research, 12*(11), 415–422.

33. Habiddin, H., Farizka, L., & bin Shuid, A. N. (2023). Chemistry students' understanding of Lewis structure, VSEPR theory, molecular geometry, and symmetry: A cross-sectional study. *Jurnal Tadris Kimiya, 8*, 1–9.

34. Hillyer, M. B., & Gibb, B. C. (2016). Molecular shape and the hydrophobic effect. *Annual Review of Physical Chemistry, 67*(1), 307–329.

35. Hosseinzadeh, B., & Ahmadi, M. (2022). Coordination geometry in metallo-supramolecular polymer networks. *Coordination Chemistry Reviews, 471*, 214733.

36. Inadomi, Y., Nakano, T., Kitaura, K., & Nagashima, U. (2002). Definition of molecular orbitals in fragment molecular orbital method. *Chemical Physics Letters, 364*(1–2), 139–143.

37. Isihara, A. (1950). Determination of molecular shape by osmotic measurement. *The Journal of Chemical Physics, 18*(11), 1446–1449.

38. Kelley, N. S., Hubbard, N. E., & Erickson, K. L. (2007). Conjugated linoleic acid isomers and cancer. *The Journal of Nutrition, 137*(12), 2599–2607.

39. Kier, L. B. (1986). Distinguishing atom differences in a molecular graph shape index. *Quantitative Structure-Activity Relationships, 5*(1), 7–12.

40. Konishi, Y., Kiyota, T., Draghici, C., Gao, J. M., Yeboah, F., Acoca, S., & Purisima, E. (2007). Molecular formula analysis by an MS/MS/MS technique to expedite dereplication of natural products. *Analytical Chemistry, 79*(3), 1187–1197.

41. Krivovichev, S. V. (2004). Combinatorial topology of salts of inorganic oxoacids: Zero-, one-, and two-dimensional units with corner-sharing between coordination polyhedra. *Crystallography Reviews, 10*(3), 185–232.

42. Kubo, R., & Tomita, K. (1954). A general theory of magnetic resonance absorption. *Journal of the Physical Society of Japan, 9*(6), 888–919.

43. Lauterbur, P. C. (1974). Magnetic resonance zeugmatography. *Pure and Applied Chemistry, 40*(1–2), 149–157.

44. Lauterbur, P. C., & Ramirez, F. (1968). Pseudorotation in trigonal-bipyramidal molecules. *Journal of the American Chemical Society, 90*(24), 6722–6726.

45. Lawton, S. L., Fuhrmeister, C. J., Haas, R. G., Jarman, C. S., & Lohmeyer, F. G. (1974). Stereochemical consequences of electron lone pairs. Examination of certain sulfosalts and chelates of trivalent arsenic, antimony, and bismuth and the crystal and molecular structure of bismuth (III) O, O'-diisopropylphosphorodithioate, $Bi[(i-C_3H_7O)_2PS_2]_3$. *Inorganic Chemistry, 13*(1), 135–143.

46. Lennard-Jones, J. E. (1949). The determination of molecular orbitals. *Proceedings of the Royal Society of London. Series A. Mathematical and Physical Sciences, 198*(1052), 1–13.

47. Malcolm, N. O., & Popelier, P. L. (2003). The full topology of the Laplacian of the electron density: Scrutinising a physical basis for the VSEPR model. *Faraday Discussions, 124*, 353–363.

48. Mashruk, S., Alnasif, A., Yu, C., Thatcher, J., Rudman, J., Peronski, L., & Valera-Medina, A. (2023). Combustion characteristics of a novel ammonia combustor equipped with stratified injection for low emissions. *Journal of Ammonia Energy, 1*(1), 21–32.

49. MÁTHÉ, K. B., & BÖRÖCZKY, K. (1999). Regular polyhedra and Hajós polyhedra. *Studia Scientiarum Mathematicarum Hungarica, 35*(3–4), 415–426.

50. McDonnell, T. E., Bail, C. A., & Tavory, I. (2017). A theory of resonance. *Sociological Theory, 35*(1), 1–14.

51. McNiel, M. E., Labbok, M. H., & Abrahams, S. W. (2010). What are the risks associated with formula feeding? A re-analysis and review. *Birth, 37*(1), 50–58.

52. Meyer, A. Y., & Richards, W. G. (1991). Similarity of molecular shape. *Journal of Computer-Aided Molecular Design, 5*, 427–439.

53. Millet, P., Johnsson, M., Pashchenko, V., Ksari, Y., Stepanov, A., & Mila, F. (2001). New copper (II)–lone electron pair elements–Oxyhalides compounds: Syntheses, crystal structures, and magnetic properties. *Solid State Ionics, 141*, 559–565.

54. Murray, R. B., Bromley, R. A., & Yoffe, A. D. (1972). The band structures of some transition metal dichalcogenides. II. Group IVA; octahedral coordination. *Journal of Physics C: Solid State Physics, 5*(7), 746.

55. Nawawi, J. P. D. H. (2021). Using simple molecular model to enhance students' understanding on molecular geometry based on VSEPR theory. *Jurnal Pendidikan dan Pembelajaran Kimia, 10*(1), 24–33.

56. Newman, R. H., & Hemmingson, J. A. (1995). Carbon-13 NMR distinction between categories of molecular order and disorder in cellulose. *Cellulose, 2*, 95–110.

57. Nicholls, A., McGaughey, G. B., Sheridan, R. P., Good, A. C., Warren, G., Mathieu, M., & Kelley, B. (2010). Molecular shape and medicinal chemistry: A perspective. *Journal of Medicinal Chemistry, 53*(10), 3862–3886.

58. Pascual, J. I., Gómez-Herrero, J., Rogero, C., Baró, A. M., Sánchez-Portal, D., Artacho, E., & Soler, J. M. (2000). Seeing molecular orbitals. *Chemical Physics Letters, 321*(1–2), 78–82.

59. Rhuland, L. E., Work, E., Denman, R. F., & Hoare, D. S. (1955). The behavior of the isomers of α, ε-diaminopimelic acid on paper chromatograms. *Journal of the American Chemical Society, 77*(18), 4844–4846.

60. Richmond, G. L. (2001). Structure and bonding of molecules at aqueous surfaces. *Annual Review of Physical Chemistry, 52*(1), 357–389.

61. Rieger, F., & Mudring, A. V. (2007). Phase transition in Tl_2TeO_3: Influence and origin of the thallium lone pair distortion. *Inorganic Chemistry, 46*(2), 446–452.

62. Roe, R. R., & Pang, Y. P. (1999). Zinc's exclusive tetrahedral coordination governed by its electronic structure. *Molecular Modeling Annual, 5*, 134–140.

63. Rohrbaugh, R. H., & Jurs, P. C. (1987). Descriptions of molecular shape applied in studies of structure/activity and structure/property relationships. *Analytica Chimica Acta, 199*, 99–109.

64. Rost, J. M., Griffin, J. C., Friedrich, B., & Herschbach, D. R. (1992). Pendular states and spectra of oriented linear molecules. *Physical Review Letters, 68*(9), 1299.

65. Schefe, J. H., Lehmann, K. E., Buschmann, I. R., Unger, T., & Funke-Kaiser, H. (2006). Quantitative real-time RT-PCR data analysis: Current concepts and the novel "gene expression's CT difference" formula. *Journal of Molecular Medicine, 84*, 901–910.

66. Schlemper, E. O., & Hamilton, W. C. (1966). The crystal structure of dimethyltin difluoride. An example of octahedral coordination of tin. *Inorganic Chemistry, 5*(6), 995–998.

67. Seburg, R. A., Patterson, E. V., Stanton, J. F., & McMahon, R. J. (1997). Structures, automerizations, and isomerizations of C3H2 isomers. *Journal of the American Chemical Society, 119*(25), 5847–5856.

68. Shibata, S., Bartell, L. S., & Gavin Jr, R. M. (1964). Molecular structure and bonding of InC_5H_5. *The Journal of Chemical Physics, 41*(3), 717–722.

69. Van Noorden, L., & Moelants, D. (1999). Resonance in the perception of musical pulse. *Journal of New Music Research, 28*(1), 43–66.

70. Vozzi, C., Negro, M., Calegari, F., Sansone, G., Nisoli, M., De Silvestri, S., & Stagira, S. (2011). Generalized molecular orbital tomography. *Nature Physics, 7*(10), 822–826.

71. Weisz, P. B. (1980). Molecular shape selective catalysis. *Pure and Applied Chemistry, 52*(9), 2091–2103.

72. Wikan, U. (1992). Beyond the words: The power of resonance. *American Ethnologist, 19*(3), 460–482.

73. Xu, X., Hu, C. L., Kong, F., Zhang, J. H., & Mao, J. G. (2011). $Ca_{10}Ge_{16}B_6O_{51}$ and $Cd_{12}Ge_{17}B_8O_{58}$: Two types of new 3D frameworks based on BO_4 tetrahedra and 1D $[Ge_4O_{12}]$ n chains. *Inorganic Chemistry, 50*(18), 8861–8868.

CHAPTER

4

Statistical Physics

LEARNING OBJECTIVES

After studying this chapter, you will be able to:

1. Understand the fundamental concepts of probability and statistics as applied to statistical physics;
2. Describe the role of quantum statistics and the density operator in statistical mechanics;
3. Explain the concept of equilibrium ensembles and their connection to thermodynamics;
4. Discuss entropy and its relation to the second law of thermodynamics;
5. Analyze ideal systems and their behavior under different temperature regimes;
6. Investigate the internal degrees of freedom of molecules in ideal systems;
7. Differentiate between high-temperature and low-temperature behavior of Fermi and Bose gases.

KEY TERMS FROM THIS CHAPTER

Probability

Quantum statistics

Equilibrium ensembles

Ideal systems

Fermi gas

Statistics

Density operator

Entropy

Internal degrees of freedom

Bose gas

UNIT INTRODUCTION

Statistical physics aims to establish a connection between the average characteristics of "macroscopic" objects, such as thermodynamic numbers, and the underlying microscopic dynamics that govern them. In the preceding chapters, we have explored how the characteristics of every molecule and atom are explained by the underlying electromagnetic forces that govern the movement of their constituent electrons and nuclei (Castellano et al., 2009). Within atoms and smaller molecules, the quantity of nuclei and electrons is typically limited to a few hundred or, in some cases, a few thousand. Large-scale objects, in contrast to tiny systems, are defined by many possible configurations. As an illustration, consider sodium-chloride crystals weighing 1 g, which is about to be added to hot water for cooking pasta. This crystal consists of an immense number of electrons, approximately $\left(3\times10^{23}\right)$, along with roughly $\left(10^{22}\right)$ Na and $\left(10^{22}\right)$ Cl nuclei. It is possible to consider a (potentially approximate) wavefunction to describe the dynamics of an extensive number of degrees for freedom. However, it is important to acknowledge that it is impossible to capture and record all the information contained in the wavefunction that describes the complete intricacies of such a system (Wang et al., 2016).

However, this restriction is not particularly detrimental, as the dynamics within a salt crystal, involving each of the motions for the electrons and nuclei, may be considered uninteresting and of limited practical value. Physicists and material researchers are fascinated by the quantifiable average macroscopic characteristics of objects and items. These include properties such as stiffness, heat and electrical conductivity, tensile strength, dielectric and magnetic susceptibilities, heat capacity, and phase transitions (Cimini et al., 2019).

4.1. INTRODUCTORY CONCEPTS

Equilibrium statistical physics utilizes mathematical methods like probability theory as well as general statistics to establish a connection amongst microscopic dynamics and macroscopic thermodynamic features (Minlos, 1968).

4.1.1. Probability and Statistics

Probability forms the foundation of statistics. The simple concept of probability aligns with the proportion of observations. As an illustration, if one rolls a die a significant number of times (N) one notices that "two" appears (N_2) times. The ratio (N_2/N) provides an approximation for the probability (P_2) of getting "two" in a single roll. Yet, one also has a preexisting notion of probability.

One will firmly maintain that $\left(P_2 = \frac{1}{6}\right)$, the probability of getting a 2 on the die is 1/6, even in the face of conflicting evidence, unless one has proof that the die is biased. The concept of probability in this context is based on a rigorous definition that involves measuring events within a defined space, with the total measure of the space being equal to unity. Here are the fundamental assumptions (Doob, 1934):

1. When considering non-intersecting sets that include (A) and (B), the probability regarding $(A \cup B)$ (i.e., how any event that occurs in (A) or (B) is realized) is equivalent to $P(A)+P(B)$;
2. When considering two spaces of events, it is possible for two sets of events (A) and (B) belonging to each space to be independent. In this case, the probability of $(A \cap B)$ (i.e., an event occurring that satisfies the criteria for both sets is equal to the product of their individual probabilities $P(A) \cdot P(B)$.

Both these properties become straightforward when the probabilities and relative frequencies of observations are properly understood (Johnson et al., 2000).

By understanding the fundamental principles of probability, one can establish a connection between the probabilities of complex events and those of simpler events. As an illustration, consider rolling two dice independently. One can calculate the probability

Remember

Remember that statistical physics provides a powerful framework for understanding the collective behavior of large systems of particles, bridging microscopic details with macroscopic observables like temperature and pressure.

of getting two ones using $\left(\frac{1}{6}\times\frac{1}{6}=\frac{1}{36}\right)$ (property 2); Similarly, one can determine the probability of receiving a four and a five $\left(\frac{1}{36}+\frac{1}{36}=\frac{1}{18}\right)$, This is because the events of getting a four upon the first die and a five on the second die, as well as getting a five on the first die and a four on the second die, are mutually exclusive (property 1) (Dekking et al., 2006).

Distributions of probability are extensively utilized in statistics. That consists of lists of probabilities assigned to sets of events that are exclusive and collectively exhaustive. As an illustration, Consider the case of rolling two dice. The possible outcomes can be divided into six equally likely possibilities, each of probability $\left(\frac{1}{36}\right)$ or into various numbers (each with a probability of $(6\times\frac{5}{2}=15$ mutually exclusive possibilities, each of probability $\left(\frac{1}{18}\right)$: By applying property 1, one can derive the distributions of $\left(P_{\text{equal}}=\frac{1}{6}\right)$, and $\left(P_{\text{different}}=\frac{5}{6}\right)$.

It is evident that the events being considered encompass the entire space, and the total sum of probabilities within a distribution equals one. Working off the distribution can be a valuable exercise.

$(P_{\text{both even}}),(P_{\text{even odd}}),(P_{\text{both odd}})$ in a similar space of events. In essence, a probability distribution associates probability with specific characteristics of the events under consideration. Some commonly used distributions in the field of statistical physics include the binomial distribution, the Poisson distribution, and the Gaussian distribution" (Dacunha-Castelle & Duflo, 2012).

Quantum mechanics, even when applied to a single atom or a small number of particles, inherently involves statistical analysis due to the probabilistic nature of observations. While the measurement involving an observable is performed upon a quantum system that is initially in a certain state, the system can be observed to be in the eigenstate related to a specific eigenvalue. The probability of obtaining this eigenstate can be calculated using the formula. The quantum state ($||i\rangle$) encompasses the full range of probability distributions linked to various operators that could potentially be used to measure the system (Suhov & Kelbert, 2014).

Statistical physics doesn't focus much on statistical interpretation. Instead, it primarily deals with the evolution of the ket out of any initial state that follows the deterministic Schrödinger Eqn. (1.7), so long that the quantum-mechanical system remains undisturbed. In contrast to that, the statistical analysis of a microscopic system aims to examine its average characteristics without requiring precise details about the initial conditions. Instead, it assumes each of the "reasonable" starting points have an equal chance of occurring. It would be ideal to link the macroscopic characteristics to time-varying averages of small observables during the evolution based on internal dynamics (Ross, 2009).

Equilibrium along with ergodicity are two fundamental assumptions that help reconcile the insignificance of these starting points from a time-average viewpoint. To achieve equilibrium, it is necessary for the system to have experienced a starting transient in the past. There should be no ongoing systematic drift, with all collective quantities fluctuating around a clearly defined average value. Ergodicity presumes that a wide range of states can be studied in a relatively short time, contrasted to the usual measurement span: Each subsequent occurrence offers unique and random outcomes from a fundamental probability distribution. An unscrupulous dice roller manipulates the initial conditions with precision, instead of relying on blind rolls to produce a series of truly random as well as independent numbers. Sometimes, an identical violation can arise in the realm of statistical physics, as demonstrated by the case involving nuclear spin via the given example. (H_2) Molecules can have an average nuclear spin of either 1 (orthohydrogen) and 0 (parahydrogen).

Both states had nearly identical energy levels, suggesting they should occur with equal likelihood. As a result, one would anticipate observing a ratio regarding ortho- to parahydrogen that aligns with their degeneracy ratio of 3:1 (Rosenkrantz, 2012). Nevertheless, the process of ortho-para inter-conversion poses a significant challenge. In typical molecular collisions, nuclear spins remain conserved, resulting in no change in the overall abundance of each species. Typically, inter-conversions take place at the walls, near magnetic impurities. Nevertheless, it is possible to store the H2 sample within a vessel that has been meticulously purged of any magnetic impurities. Consequently, a peculiar proportion of ortho- towards parahydrogen quantity can be maintained for a prolonged period. On the other hand, in the presence of magnetic impurities, the nuclear spins undergo haphazard rearrangement, leading to the restoration of the predicted 3:1 equilibrium distribution and ensuring the system's ergodicity. Simply put, the ergodic hypothesis suggests that the system remains random enough that nothing preserved resources hinder access to a significant subset for the state space during short periods when compared with the experiment's duration. When a system reaches a state for equilibrium and ergodicity, it is possible to replace time averages with averages over the distribution of probabilities for the microscopic states. Let us outline the typical mathematical framework utilized to explain the erratic distribution for quantum states" (Miller & Miller, 2019).

4.1.2. Quantum Statistics and the Density Operator

One aims to establish a formal understanding of our limited knowledge regarding the exact quantum condition of a system, while accurately characterizing its statistical properties. A system can exist in various quantum states, $(|a_1\rangle), (|a_2\rangle)$, each with its own probability of $(w_1, w_2, \ldots)$ occurrence.

The states $(|a_1\rangle, (|a_2\rangle)$, may not necessarily be eigenstates that are observable, which might not be orthogonal, and their number could potentially surpass the dimension for the Hilbert area of states (Fano, 1957). Ensuring the proper normalizing of probability $(\sum_i w_i = 1)$

Typically, a measurement of a single observable (B) on including a system should yield:

$$[B] = \sum_i w_i \langle a_i|B|a_i\rangle = \sum_i w_i \sum_b b\,|\langle a_i|b\rangle|^2 \quad (4.1)$$

where; the value of$(\{|b\rangle\})$ serves as the foundation for the eigenkets of (B) (B), each with its corresponding eigenvalues (b). One uses brackets of squares () to represent the group's average, which refers to the statistical mean among the quantum medians. The mean values are calculated utilizing the density operator, which is a key tool in analyzing systems (Man'ko et al., 2020).

$$\hat{\rho} = (\sum_i w_i\, |a_i\rangle\langle a_i|) \quad (4.2)$$

where;

$$[B] = \sum_i w_i \langle a_i|B|a_i\rangle = \sum_i w_i \sum_b b\,\langle b|a_i\rangle\langle a_i|b\rangle = \sum_b b\,\langle b|(\sum_i w_i\,|a_i\rangle\langle a_i|\,)|b\rangle = \sum_b b\,\langle b|\hat{\rho}|b\rangle = \mathrm{Tr}(\rho B) \quad (4.3)$$

The density operator encompasses all dynamic and statistical characteristics of the system. The operator $(\hat{n})$ possesses the property of self-adjointness, allowing for its diagonalization. The density operator can be expressed on its diagonal basis (Tanaka & Komaki, 2005).

$$\hat{\tilde{n}} = \sum_i P_m \,|\, P_m P_m \,| \quad (4.4)$$

where; the kets $(|\tilde{n}_m\rangle)$ make up the full orthonormal base of the Hilbert space, which is different from the (in contrast to the $(|a_i\rangle)$ mentioned in the definition. These kets correspond to positive eigenvalues (P_m). Take note of the standardization (Ishihara, 2021).

$$\mathrm{Tr}(\hat{\rho}) = \sum_b \langle b|\hat{\rho}|b\rangle = \sum_b \sum_i w_i \langle b|a_i\rangle\langle a_i|b\rangle = \sum_i w_i \sum_b \langle a_i|b\rangle\langle b|a_i\rangle = \sum_i w_i \langle a_i|a_i\rangle = \quad (4.5)$$

Furthermore, it is worth noting that when $(\sum_m P_m = 1)$:, the eigenvalues (P_m)may prove interpreted to be probabilities.

In the specific scenario where a statistical worker represents a solitary pure state $(\hat{}\{\rho\} = |a\_1\rangle\langle a\_1|)$ it reverts to the conventional deterministic quantum mechanics (Hofmann, 2020).

$$[B] = \mathrm{Tr}(\hat{\rho}B) = \langle a_1|B|a_1\rangle = \langle B\rangle.$$

In this scenario (and only in that scenario), $(\hat{\rho})$ exhibits the characteristics of a projector $(\hat{\rho}^2 = \hat{\rho})$.

4.2. EQUILIBRIUM ENSEMBLES

Now think about a large, self-contained system (referred to as the "universe") (U) with a described energy (microcanonical ensemble) that has some level of uncertainty (ΔE). According to the ergodic hypothesis, every quantum state for (U) is equally probable, which leads to the basic postulate for equilibrium numerical physics (Eu, 2013).

Put simply, the likelihood of a particular state $(|i\rangle)$ occurring is determined by constant $(P = 1/\Omega_U(E;\Delta E))$ when the energy $(E_U) of (|i\rangle)$ falls within the range $(E \leq E_U \leq E + \Delta E)$. Otherwise, the likelihood is zero. The value of $(\Omega_U(E;\Delta E))$represents the count of distinct states of (U)within the specified energy range $(E \leq E_U \leq E + \Delta E)$. By following this approach, the probability$\left(P_U^i\right)$ is appropriately normalized to a value of one. Based on this fundamental principle, one could deduce the probability distribution for a system (S) that is in thermal equilibrium about the rest of the universe (W). The overall Hamiltonian $(E_U \approx E_S + E_W)$is presumed to include a negligible interaction term $(H_{SW} \approx 0)$, resulting in particular issue: determining the likelihood (P_m) of observing system (S) in a specific quantum state $(|m)$ of energy (E_m)? Given the weak coupling within (S) and (W) one can assume how the states within (S) and (W) vary independently and randomly. The only connections among (S) along with (W) are caused by the preservation of energy (Schwabl, 2006). As a result;

$$P_U = P_S^m P_W \qquad (4.6)$$

The microcanonical distributions play a crucial role in both (U) and (W) independently. The occurrence of (U)–states is governed by the probability $(PU{=}1/\Omega UE;\ \Delta E)$, while in (W), each state has a probability of $(PW{=}1/\Omega WE{-}Em;\ \Delta E)$. Here, the residual energy for (W) is calculated by subtracting the energy (Em) on (S) from the total energy (E). By utilizing Eqn. (4.6), one can derive the probability distribution for (S):

$$P_S^m = \frac{P_U P_W}{\Omega_U(E;\Delta E)} = \frac{\Omega_W(E-E_m;\Delta E)}{\Omega_U(E;\Delta E)}. \qquad (4.7)$$

It is worth noting that the energy (E_m) within a system (S) is typically a minuscule portion of the overall energy from the universe (E) for the majority of (U)–states. It is possible to expand the denominator (specifically, its logarithm) around (E) (Luzzi et al., 2002).

$$\ln \Omega_W (E - E_m; \Delta E) = \ln \Omega_W (E; \Delta E) - \beta E_m + \cdots, \quad (4.8)$$

The slope $(\beta = \frac{\partial \ln \Omega_W(E;\Delta E)}{\partial E}\Big|_{E=E})$ has dimensions that correspond to inverse energy. The linear proximity of the logarithm is highly accurate when the energy of the system is much smaller than the energy scale

$$(\beta = \frac{\partial \ln \Omega_W(E;\Delta E)}{\partial E}\Big|_{E=E}).$$

After substituting it back into Eqn. (4.7), one may be able to obtain:

$$P_S^m = \frac{\Omega_W(E;\Delta E)}{\Omega_U(E;\Delta E)} e^{-\beta E_m}, \quad (4.9)$$

This is an incredibly impressive outcome for a least two notable reasons (Zurek, 2018):

i. The entire reliance on the state $(|m\rangle)$ of (S) is limited to the exponential component since the normalization aspect $(Z^{-1} = \frac{\Omega_W(E;\Delta E)}{\Omega_U(E;\Delta E)})$ remains unaffected by $(|m\rangle)$

ii. The likelihood of a state $(|m\rangle)$ of (S) depends solely on its energy (E_m) following a straightforward exponential relationship.

One can rewrite Eqn. (4.9) by including the normalizing factor (Z^{-1}) along with removing the label (S):

$$P_m = P(E_m) = \frac{e^{-\beta E_m}}{Z}. \quad (4.10)$$

Eqn. (4.10) illustrates the Boltzmann equilibrium distribution of probabilities for each state of a generic system that is in weakened thermal contact within a vast environment. This distribution is related to the Gibbs canonical ensemble. Since the distribution of probabilities (4.10) is normalized, a substitute expression for (Z) can be derived solely based on the properties of the system (S) being studied (Ilievski et al., 2017).

$$Z = \sum_m e^{-\beta E_m} = \mathrm{Tr}\left(e^{-\beta H}\right), \quad (4.11)$$

Without any mention of (U) and (W), one can instead use (H) instead of (H_s). The partition function, denoted as (Z), is a commonly used term in the field of statistical physics. It is important to note the fact the sum in Eqn. (4.11) encompasses all microstates, such as all components for degenerate levels with the same energy. Another way to express the identical sum is:

$$Z = \sum_E n_{\text{deg}}(E) e^{-\beta E} \quad (4.12)$$

where; $(n_{\text{deg}}(E))$ represents the count of degenerate states associated with the energy level (E). It is worth noting that each trace within Eqn. (4.11) may be evaluated using any convenient basis. The density operator linked by the Boltzmann equilibrium distribution of probabilities exhibits a diagonal pattern when represented in terms of energy:

$$\hat{\rho}_{eq} = \sum_m \frac{e^{-\beta E_m}}{Z} |m\rangle\langle m| = \frac{1}{Z} e^{-\beta H}. \quad (4.13)$$

By using a suitable unified transformation, it is possible to express the density operator upon any basis (De Ninno & Fanelli, 2012).

4.2.1. Connection to Thermodynamics

Consider a scenario where the system (S) can be divided by the subsystems ( ) along with $(S_2)\left(\left(E_{m_1m_2} \approx E_{1m_1} + E_{2m_2}\right)\right)$. While thermal equilibrium is reached among $(S_1),(S_2)$, while the surrounding environment(W), the pattern of distribution:

$$P_{m_1m_2} = Z^{-1}e^{-\hat{a}E_{m_1m_2}} = \frac{e^{-\hat{a}\left(E_{1m_1}+E_{2m_2}\right)}}{\sum_{n_1n_2} e^{-\hat{a}\left(E_{1n_1}+E_{2n_2}\right)}},$$

Might be factorize into:

$$P_{m_1m_2} = \frac{e^{-\hat{a}E_{1m_1}}}{\sum_{n_1} e^{-\hat{a}E_{1n_1}}} \cdot \frac{e^{-\hat{a}E_{2m_2}}}{\sum_{n_2} e^{-\hat{a}E_{2n_2}}} = \frac{e^{-\hat{a}E_{1m_1}}}{Z_1} \cdot \frac{e^{-\hat{a}E_{2m_2}}}{Z_2} \quad (4.14)$$

The decomposition reveals which at equilibrium, (S_1) and (S_2) adhere to separate Boltzmann distributions, $(P_{S_i m_i} = \frac{e^{-\beta E_{i m_i}}}{Z_i})$, and have an identical (β) parameter. It can be inferred from this observation that the amount of intensive quantity (U) could potentially depend on temperature (T).

The internal force energy (U) within thermodynamics is typically defined as the mean for the energy operator, following the principles of statistical physics (Sommerfeld, 2012).

$$U = [H] = \mathrm{Tr}(\hat{\rho}H) = \sum_m E_m P_m = \frac{1}{Z}\sum_m E_m e^{-\beta E_m}. \quad (4.15)$$

When one derives (U) considering (β), one finds that it results in the negative of the squared energy fluctuation.

$$\frac{\partial U}{\partial \beta} = -Z\sum_m E_m^2 e^{-\beta E_m} + \left(\sum_m E_m e^{-\beta E_m}\right)^2$$

$$= -\langle H^2\rangle - \langle H\rangle^2 = -\langle (H - \langle H\rangle)^2\rangle \quad (4.16)$$

It can be observed that the internal energy reduces as the value of $(\hat{a})$ increases, as indicated by $(\frac{\partial U}{\partial \beta} \leq 0)$.. It can be inferred from this observation that there may be an inverse relationship between (β) and temperature. A detailed outline of the exact correlation is provided below.

Firstly, it is important to consider that (Z) behaves as a function of multiplicity ([Z = Z_1 Z_2 "{over a system made up of two subsystems, see Eqn. (4.14), Consequently, its logarithm can be treated as an additive function $((\ln Z = \ln Z_1 + \ln Z_2))$.The representation of .$(\ln Z)$ as an extensive thermal quantity is essential in this context. Provide a clear definition of the function (Cahn et al., 1997).

$$F = -\frac{\ln Z}{\beta}. \quad (4.17)$$

Using this definition, we derive a relation between the (β)derivative of the extensive quantity $(\ln Z)$ and the internal energy (U):

By employing this definition, a connection can be established within the (β)derivative corresponding to the extensive amount $(\ln Z)$ along with the internal energy (U):

$$\left.\frac{\partial(\beta F)}{\partial\beta}\right|_{V,N} = -\frac{\partial(\ln Z)}{\partial\beta} = -\frac{1}{Z}\frac{\partial Z}{\partial\beta} = -\frac{\partial}{\partial\beta}\left(\sum_m \frac{e^{-\beta E_m}}{Z}\right) = -\sum_m \frac{\partial}{\partial\beta}\left(\frac{e^{-\beta E_m}}{Z}\right) = \sum_m E_m e^{-\beta E_m} = [H] = U. \tag{4.18}$$

Specifically, the average energy, defined within Eqn. (4.15). Eqn. (4.18) establishes a connection between the derivative for (βF) concerning (β) and the internal energy. This identification reminds us of a well-known thermodynamic identity that relates to free energy or temperature: Beginning with the fundamental definition of energy that is free $(F=U-TS)$ (in which (S) represents entropy), it is possible to derive:

$$U = F + TS = F - T\frac{\partial F}{\partial T} = -T^2\left(-T^{-2}F + T^{-1}\frac{\partial F}{\partial T}\right) = -T^2\frac{\partial(T^{-1}F)}{\partial T}. \tag{4.19}$$

This expression may be written more concisely by utilizing the derivative of its opposite temperature:

$$\frac{\partial(T\ \ F)}{\partial(\ \ )} \tag{4.20}$$

In Eqns. (4.19) and (4.20), the derivatives with respect to temperature (T) are taken while keeping the number of particles (N) and volume (V) constant. Upon examining Eqns. (4.18) and (4.20), it becomes evident that:

i. It is quite reasonable to equate the vast quantity (F)stated statistically within Eqn. (4.17) using the free energy (F) for thermodynamics (Cole, 2002).

ii. (β) must therefore be directly related to the reciprocal of the temperature (T^{-1}).

The relationship between (T) and (β^{-1}) energy and temperature is represented by the Boltzmann constant (k_B), It acts as a conversion factor within temperature in Kelvin [K] and energy in Joules [J]:

$$\beta = \frac{1}{k_B T}. \tag{4.21}$$

By comparing Eqn. (4.21) to the empirical definition given above using the ideal-gas thermometer, one can establish the relation $\left(k_B = \frac{R}{N_A}\right)$ involving the gas \(R\) The resulting value is 1.38065×10−23 J/K, which is equivalent to 86.1733 µeV/K. Temperature is commonly represented in statistical physics through an energy combination $(k_B = 1.38065 \times 10^{-23}\,\text{J/K} = 86.1733\,\mu\text{eV/K})$ and the (β) notation (Buxton, 2021).

Now one can understand the significance of temperature: the likelihood of different microstates in a system at thermal equilibrium is solely determined by their energies.

This relationship is quite simple: for a given temperature (T), for a given temperature (T), the probability $(P(E_m))$ of a specific state with energy (E_m) correlates to $(\exp(-\beta E_m))$ (Curado & Tsallis, 1991).

It is important to consider that the normalized factor (Z^{-1}) in Eqn. (4.10) affects the state of equilibrium probability (P_m) for a specific energy eigenstate $(|m)$. in the system. Nevertheless, the probability ratio $(P_m/P_n = \exp[\beta(E_n - E_m)])$solely relies on the energy distinction among two states $(|m)$ and $(|n)$ (Krekelberg et al., 2013).

The mentioned equations, specifically Eqns. (4.17) along with (4.21), establish a clear connection between the principles of microscopic dynamics and observed average characteristics at a macroscopic level. This link bridges the gap between statistics and thermodynamics. Using Eqn. (4.17), it is possible to expand on this relationship and establish a direct connection between the partition function \(Z\) and all the thermodynamic features of a system in equilibrium. Here, we provide the formulas for various important thermodynamic quantities, both extensive and intensive (Collell & Fauquet, 2015):

$$F = -k_B T \ln Z = -\frac{\ln Z}{\beta} \quad (4.22)$$

$$U = F + TS = \frac{\partial(\beta F)}{\partial\beta}|_{V,N} = -\frac{\partial(\ln Z)}{\partial\beta}|_{V,N} \quad (4.23)$$

$$S = -\frac{\partial F}{\partial T}|_{V,N} = k_B T\frac{\partial \ln Z}{\partial T} + k_B \ln Z = \frac{U}{T} + k_B \ln Z \quad (4.24)$$

$$C_V = \frac{\partial U}{\partial T}|_{V,N} = -k_B\beta^2\frac{\partial U}{\partial\beta}|_{V,N} = k_B\beta^2\frac{\partial^2 \ln Z}{\partial\beta^2} \quad (4.25)$$

$$]P = -\frac{\partial F}{\partial V}|_{V,N} = \left(\frac{N}{V}\right)^2\frac{\partial(F/N)}{\partial(N/V)}|_{V,N} \quad (4.26)$$

$$G = F + PV \quad (4.27)$$

$$H = F + PV + TS \quad (4.28)$$

4.2.2. Entropy and the Second Principle

The relationship between entropy along statistics (Eqn. 4.24) is only valid when the system is in equilibrium. There exists a comprehensive definition of entropy that can be applied to any kind of statistical distribution for quantum states $(\hat{\rho})$, regardless of whether it belongs to an equilibrium group or not.

$$S = -k_B \mathrm{Tr}(\hat{\rho}\ln\hat{\rho}). \quad (4.29)$$

This definition aligns with the intuitive concept of entropy to be a quantification of chaos. Let's look at the basis $(|\rho_m\rangle)$whereas $(\hat{\rho})$ is diagonal. Here, (`{ρ} = ∑_m P_m|ρ_m⟩⟨ρ_m|). . On the basis (Ozawa et al., 2003):

$$S = -k_B \sum_m P_m \ln P_m. \quad (4.30)$$

It is possible to calculate (S) for a particularly ordered distribution, which is a pure state (i.e., $(Pm=1)$ over a single state (m) along with 0 for all others). In this case, we find that $(S=0)$ as every term disappears. The increase in (S) occurs when multiple states possess non-zero probabilities. In the case of an entirely random distribution where each state has an equal probability $Pm=1$, one can derive the following result.

$$S = -k_B \sum_{m=1}^{\Omega} \Omega^{-1} \ln(\Omega^{-1}) = k_B \ln(\Omega).$$

It can be inferred that entropy represents a logarithmic measure for the multitude of variables that the system can access.

Did you know?

Did you know that entropy, a measure of disorder or randomness in a system, increases over time as stated by the Second Law of Thermodynamics? This principle underpins the direction of natural processes towards equilibrium.

At equilibrium, the concept of equilibrium in statistical physics aligns with Eqn. 4.24. One can easily confirm this by substituting an equilibrium density operator $(\widehat{\rho_{eq}})$ from Eqn. 4.13 into the general definition (Kapur & Kesavan, 1992).

$$S = -k_B \sum_m P_m \ln P_m = -k_B \sum_m \frac{e^{-\beta E_m}}{Z} \ln\left(\frac{e^{-\beta E_m}}{Z}\right)$$

$$= -k_B \sum_m \frac{e^{-\beta E_m}}{Z}(-\beta E_m - \ln Z)$$

$$= k_B \sum_m \frac{e^{-\beta E_m}}{Z}(\beta E_m + \ln Z)$$

$$= k_B \beta [H] + k_B \frac{\ln Z}{Z} \sum_m e^{-\beta E_m}$$

$$= \frac{U}{T} + k_B \ln Z. \tag{4.31}$$

In addition, it can be demonstrated that when the density matrix $(\hat{\rho})$matches the equilibrium density matrix $(\widehat{\rho_{eq}})$described by the Gibbs-Boltzmann equation (Eqn. 4.13), the system entropy, defined in Eqn. 4.29, reaches its maximum value while adhering to the constraints of assigned internal energy $(\mathrm{Tr}(\hat{\rho}H) = U)$,, $volume(V)$, and number of particles (N) (Martyushev & Seleznev, 2006). The entropy of any given density operator (ρgen), which follows the constraints upon (U),(V), and (N), is always lower than or equal to the entropy that defines the equilibrium distribution $(\widehat{\rho_{eq}})$ (Eqn. 4.13). Every isolated system naturally tends to move toward a state of equilibrium; as a result, this result, derived from the principle for maximum entropy, demonstrates the congruence between statistical physics and the second law of thermodynamics (Wang, 2017). During a spontaneous change towards equilibrium $(\widehat{\rho_{gen}} \rightarrow \widehat{\rho_{eq}})$,, the entropy rises (S_{gen}) up to (S_{eq}) from its initial value to its final value. This result can be applied to situations where the values of (U),(V), or (N) change. However, for our purposes, it is enough to acknowledge that the observed increase in entropy as a system approaches equilibrium is based on statistical analysis (He & Kang, 2010).

4.3. IDEAL SYSTEMS

Before delving into the intricate characteristics of complex objects like a bicycle and a vessel of soup, researchers have wisely opted to study simpler systems. These systems typically consist of uniform collections of atoms or molecules, either of the same type or a small number of types (Balcerzak et al., 2007). These simpler systems make up a wide range of materials. The reason for investigating materials is that by examining the properties of a complex organized object, one can gain insight into how its individual components function (Leff, 1969). These components, in turn, are influenced by their shapes along with materials. As an illustration, the overall heat ability of a bicycle was nearly the same as the combined heat capacities of its components. When examining the characteristics of a material on a large scale, researchers often focus on extremely wide samples. This allows them to make certain assumptions in their analysis, such as the interactions using the surroundings being relatively insignificant. Considering the interaction between surface atoms/molecules and the environment, it is observed that a layer of approximately $(\sim 1\,\text{nm})$ thickness is involved. For samples with linear dimensions larger than, (~ 1 μm) the error caused by this estimate is unlikely to be overly significant (Esposito et al., 2009).

Despite assuming a perfectly uniform bulk sample, the method:

1. Calculate the energy spectrum (E_m) and corresponding eigenstates $(\hat{\rho})$ for the system.

2. Calculate the division function (Z) at a specific temperature $(T)(or(\beta = 1/(k_B T)))$, using Eqn. (4.11).

3. Use Eqn. (4.13) to generate the state of equilibrium density operator $(\hat{n})$.
4. Calculate the average values of macroscopic quantities using the expression $([B] = \mathrm{Tr}(\hat{\rho}B))$, as shown in Eqn. (4.3) (Pérez & Sauer, 2010);

It is not easily applicable to any practical system, primarily because of the challenges involved in step 1. Fortunately, a satisfactory conclusion can be reached for a few ideal systems through a program of statistical physics. When we refer to ideal systems, we are talking about systems made up of distinct parts (particles) that have very little interaction with each other. Ideal systems possess (Blanciforti & Green, 1983):

$$H = \sum_{i=1}^{N} H_i, \qquad (4.32)$$

Here, the dynamics for a small set of degrees of freedom, such as the position along with the spin of one particle, are governed by (H_i). By utilizing statistical methods, we can accurately predict the thermodynamic features of ideal systems due to their inherent simplicity.

There is no perfect system within nature, but photons and neutrinos come close to being non-interacting particles. If such a system were to exist, it would not be possible to apply the methods for equilibrium statistics to it. This is because the system would rather be ergodic and therefore would never reach a state of equilibrium. Nevertheless, there are certain characteristics of systems with weak interactions that bear a resemblance to those of ideal systems. Understanding the attributes of gaseous content can be achieved by analyzing the characteristics of ideal systems, both qualitatively and quantitatively. It is important to note that the applicability of the information provided here is limited and there may be a risk of artifacts because of the idealized characteristics of non-interacting systems. A thorough comprehension of the thermodynamic characteristics of real materials often necessitates the utilization of more advanced techniques to consider interactions (Sands & Dunning-Davies, 2014).

Imagine a perfect system made up of a single kind of identical particles that don't interact with each other, such as electrons, atoms, or molecules. The fundamental basis for states is a factorized basis, as shown in Equations (2.60) and (2.61), in which the quantum numbers (α_i) are used to figure out the status $(|\alpha_i\rangle)$ for particles (i). Setting aside the permutational symmetry demands, the partition function

$$Z = \sum_{\alpha_1 \alpha_2 \cdots \alpha_N} \exp[-\beta(E_{\alpha_1} + E_{\alpha_2} + \cdots + E_{\alpha_N})]$$

$$= \sum\nolimits_{\alpha_1 \alpha_2 \cdots \alpha_N} e^{-\beta E_{\alpha_1}} e^{-\beta E_{\alpha_2}} \cdots e^{-\beta E_{\alpha_N}}, \qquad (4.33)$$

When dealing with bosons, it's worth noting that there can be instances where certain quantum numbers (α_i)'s may coincide, resulting in a smaller number of permutations. To calculate the number of permutations, start by arranging the single-particle states $(|\alpha\rangle)$, in a specific order, such as increasing energy, like $(|0\rangle), (|1\rangle), (|2\rangle)$, and so on. Then tally up $(n_0, n_1, n_2, \ldots)$ the occurrences of each individual particle state in every term for the sum:

$$\sum\nolimits_{\alpha_1 \alpha_2 \cdots \alpha_N} e^{-\beta(E_{\alpha_1} + E_{\alpha_2} + \cdots + E_{\alpha_N})}.$$

It is evident that the total of those occupation numbers equals $(n_0 + n_1 + n_2 + \cdots = N)$.. The total number of permutations for a given set of indices $(\alpha_1, \alpha_2, \ldots, \alpha_N)$can be calculated using the formula $(N!/(n_0! n_1! n_2! \cdots))$.. This expression is applicable to fermions as well, as it holds true for all $(n_å != 0! = 1) or (1! = 1)$ Now, one can adjust the function of partitions to eliminate any duplicate counts of identical states.

$$Z = \sum\nolimits_{\alpha_1 \alpha_2 \cdots \alpha_N} \frac{1}{n_0! n_1! n_2! \ldots} \exp\left[-\beta \sum\nolimits_{i=1}^{N} E_{\alpha_i}\right]. \qquad (4.34)$$

The Eqn. (4.34) considers all potential (α_i)'s over bosons and respects Pauli's principle by accounting for different (α_i)' respecting in the scenario of fermions (Chapman et al., 2016).

4.3.1. The High-Temperature Limit

When temperatures are low, the Boltzmann exponential aspect gives preference to states via a total energy $(\sum_{i=1}^{N} E_{\alpha_i} = \sum_\alpha n_\alpha E_\alpha)$ that is as small to be possible. This means that only a small number of low-energy single-particle contents are occupied. One will revisit this particular temperature range in Section 4.3.2 (Mézard & Parisi, 2000).

On the other hand, when temperatures are high (small (β)), and it is generally assumed that the single-particle range has no upper limit, the occupied probabilities of the vast number $((N))$ for single-particle contends in energy $(\sim k_B T)$ are remarkably alike. The likelihood of one state being occupied is quite low. States with distinct quantum numbers for each of the (N)–particles are significantly more abundant compared to states where two or more of the quantum numbers are the same (Blatter & Ivlev, 1994).

Tip

In the high-temperature limit, thermal energy dominates, and classical statistical mechanics can be used to describe the behavior of particles. This simplification is especially useful for understanding gases and other dilute systems.

Consequently, at elevated temperatures, the dominant factor in the boson partition operates arises from states with $(n_\alpha = 0)$ or (1) at most. The terms in the total of Eqn. (4.34) with $(n_\alpha > 1)$ make an insignificant contribution to the overall value of (Z) Therefore, in Eqn. (4.34), we make an approximation by assuming $(n_0! n_1! n_2! \cdots \approx 1)$ Even for the fermionic case, we consider the unphysical regard to with $(n_\alpha > 1)$ within the sum. Therefore, we derive a rough partition function for both bosons and fermions at high temperatures (Lee, 2007).

$$Z \approx \frac{1}{N!} \sum_{\alpha_1 \alpha_2 \cdots \alpha_N} \exp\left(-\beta \sum_{i=1}^{N} E_{\alpha_i}\right) = \frac{1}{N!} Z_1 \cdot Z_2 \cdot \cdots \cdot Z_N = \frac{(Z_1)^N}{N!}. \quad (4.35)$$

In this context, the single-particle divide function $(\exp(-\beta E_\alpha))$ calculates the sum of the Boltzmann weighs $(\exp(-\beta E_\alpha))$ across all one-particle contends $(|\alpha\rangle)$ It is important to note that (Z_1) is not to be mistaken for the complete (Z) of Eqn. (4.11), which encompasses the states of the whole system $(|m\rangle)$ instead (Algin & Senay, 2012).

At high temperatures, the free energy can be expressed as:

$$F = -\frac{\ln Z}{\beta} \approx -k_B T \ln\left(\frac{(Z_1)^N}{N!}\right) \approx -k_B T \left(N \ln Z_1 - N \ln \frac{N}{e}\right) = -N k_B T \ln\left(\frac{e Z_1}{N}\right), \quad (4.36)$$

In our analysis (DeWitt, 1966), one utilized a Stirling approximation $(\ln(N!) \approx N \ln(N/e))$, that proves to be highly precise when dealing with large values of (N).

4.3.2. Internal Degrees of Freedom of Molecules

In this discussion, one will briefly explore an instance of diatomic molecules, with the assumption that rotational along vibrational motions are independent. The partition function can be written as $(Z_{1,\text{vib}} \cdot Z_{1,\text{rot}})$ which allows us to separate the free energy $(F_{1,\text{int}} = -k_B T \ln Z_{1,\text{int}})$ into terms related to rotation and vibration (Stratt, 1984).

In the realm of rotational statistics, the function of partition can be expressed as:

$$Z_{1,\mathrm{rot}} = \sum_{l=0}^{\infty}(2l+1)\exp\left(-\frac{\Theta_{\mathrm{rot}}}{T}l(l+1)\right),$$

where, $(\Theta_{\mathrm{rot}} = \frac{\hbar^2}{2Ik_B})$ represents the trait's rotational temperatures. It is generally not possible to find a closed-form expression for the series (4.37). However, when large $(\Theta_{\mathrm{rot}}/T)$,is large, it may be calculated using a simple integral (Schirmer, 2000).

$$Z_{1,\mathrm{rot}} \approx \frac{T}{\Theta_{\mathrm{rot}}}. \quad (4.38)$$

The thermal functions are influenced by the rotational contributions at high temperatures:

$$U_{1,\mathrm{rot}} = k_B T, \quad (4.39)$$

$$U_{1,\mathrm{rot}} = k_B T, \quad (4.40)$$

$$C_{V,1,\mathrm{rot}} = k_B, \quad (4.41)$$

$$S_{1,\mathrm{rot}} = k_B\left(1+\ln\left(\frac{T}{\Theta_{\mathrm{rot}}}\right)\right). \quad (4.42)$$

When temperatures are lower $(T \sim \Theta_{\mathrm{rot}})$, it is more accurate to approximate $(Z_{1,\mathrm{rot}})$. the series (4.37) by truncating it to a limited number of terms $(l_{\max} \sim \sqrt{T/\Theta_{\mathrm{rot}}})$. As the temperature rises, the rotational level of freedom becomes more active, reaching the traditional equipartition limit at high temperature $(T \gg \Theta_{\mathrm{rot}})$. The population sizes of relative equilibrium $(P_l = (2l+1)\exp\left(-\frac{l(l+1)\Theta_{\mathrm{rot}}}{T}\right)/Z_{1,\mathrm{rot}}\backslash)$ explain the intensities from rotational structures within molecular spectra (Pople, 1954). In the realm of vibrational statistics within harmonic estimation, the series can be evaluated to yield a closed form. This closed form offers a precise partition function that remains valid across all temperatures:

$$Z_{1,\mathrm{vib}} = \frac{1}{2\sinh(\Theta_{\mathrm{vib}}/2T)}, \quad \text{where} \quad \Theta_{\mathrm{vib}} = \frac{\hbar\omega}{k_B}. \quad (4.43)$$

The thermodynamic properties associated with the vibrational degrees of freedom are (Lyubartsev & Laaksonen, 2004):

$$F_{1,\mathrm{vib}} = \frac{\hbar\omega}{2} + k_B T\ln(1-e^{-\Theta_{\mathrm{vib}}/T}), \quad (4.44)$$

$$U_{1,\mathrm{vib}} = \frac{\hbar\omega}{2} + \hbar\omega\frac{e^{-\Theta_{\mathrm{vib}}/T}}{1-e^{-\Theta_{\mathrm{vib}}/T}}, \quad (4.45)$$

$$C_{V,1,\mathrm{vib}} = k_B\left(\frac{\Theta_{\mathrm{vib}}}{T}\right)^2\frac{e^{\Theta_{\mathrm{vib}}/T}}{(e^{\Theta_{\mathrm{vib}}/T}-1)^2}, \quad (4.46)$$

$$S_{1,\mathrm{vib}} = k_B\left(\ln\left(1-e^{-\Theta_{\mathrm{vib}}/T}\right)+\frac{\Theta_{\mathrm{vib}}}{T}\frac{e^{-\Theta_{\mathrm{vib}}/T}}{1-e^{-\Theta_{\mathrm{vib}}/T}}\right). \quad (4.47)$$

Similar to rotations, the vibrating degree freedom also becomes active as temperature rises, hitting the classical limit when the temperature is much higher than the vibrational temperature $(T \gg \Theta_{\mathrm{vib}})$. It should be noted that the upper-temperature limit of one oscillator adds $(k_B T) to (U_1)$ (and thus (C_{V, 1})), which is equal to the impact of the two rotating degrees of freedom. This occurs because a single harmonic oscillator is linked to two quadratic components (kinetic and potential) within the Hamiltonian, similar to the two quadratic terms for rotating degrees of freedom that describe the motion (Shah & Maginn, 2011).

4.3.3. Low-Temperature Fermi and Bose Gases

It has been observed that at elevated temperatures, non-interacting identical fermions and bosons exhibit behavior akin to an ideal classical gas, with the inclusion of supplementary contribution to internal degrees of freedom. Nevertheless, when temperatures drop, noticeable disparities arise among fermions and bosons as a result of their unique limitations on the occupation of single-particle states.

By shifting the emphasis from particles (designated by (i)) towards single-particle states (designated by (α)). the precise divide function (Z) (Eqn. (4.34)) during random temperature (T) may be computed. The averages over the individual quantum values (α_i) turn into sums across the position's numbers (n_α), for the individual states. By representing the total energy of N particles in terms of $(n_{\acute{a}})$, where the sum of individual energies $(\sum_i E_{\alpha_i})$ grows $(\sum_\alpha n_\alpha E_\alpha)$, the rapid factorizes as:

$$Z = \sum_{\{n_\alpha\}} \prod_\alpha (\exp(-\beta E_\alpha))^{n_\alpha}.$$

Next, one can determine the partition function (Z):

$$Z = \sum_{\{n_\alpha\}} \prod_\alpha (\exp(-\beta E_\alpha))^{n_\alpha}. \tag{4.48}$$

In this context, (n_α)denotes the occupation numbers: either (0) or (1) for fermions (in accordance with the Pauli exclusion principles) and $(0,1,2,3,\ldots)$ for bosons. The coefficients of binomials that account for overcounting within the sum across single-particle contends (as shown in Eqn. (4.34)) are conveniently managed in a sum over (n_α)guaranteeing that each N-particle state (identified by the position numbers as well) is accurately counted only once (Cowan, 2019).

Nevertheless, calculating this amount over $(\{n_{\acute{a}}\})$ with the fixed total amount of particles ( ) limitation is exceptionally difficult. To address this, one may construct the grand canonical group, which allows for fluctuations in (N) This ensemble explains a system that has a limited exchange of atoms using the surroundings. Figure 4.11 demonstrates the necessary modification to the energy eigenvalue within the Eqn. (4.11) of the divide function, in which (μN) was subtracted.

When considering a collection of identical particles, the ultimate partition function $(\mathcal{Q})$ can be obtained by adding the standard partition function (Z) across all possible (N) values (Pathria, 1998):

$$\mathcal{Q} = \sum_{N=0}^{\infty} \Omega(N) \exp(-\beta(E_M(N) - \mu N)) = \mathrm{Tr}(e^{-\beta(H-\mu\hat{N})}) \tag{4.49}$$

In this context, $(\Omega(N))$refers to the multiplicity of states using (N) particles, while$(E_M(N))$ reflects the associated energy eigenvalue. The overall Hamiltonian for the system is denoted as (H) and $(\hat{N})$is the number of operators. The grand potential (similar to Eqn. (4.22)) establishes a link to the laws of thermodynamic:

$$J(T,V,\mu) = PV(T,\mu) = k_B T \ln \mathcal{Q}\,(T,V,\mu). \tag{4.50}$$

The equations provided (Eqns. (4.81)–(4.86)) allow for the calculation of various thermodynamic parameters over a system that is in equilibrium using energy and particle reservoirs.

Equipped with this powerful tool, one embarks on the calculation of the grand division function $(\mathcal{Q})$ over a system consisting of non-interacting comparable bosons or fermions. Considering that (N) is the sum of $(N = \sum_\alpha n_\alpha)$, one can rearrange the exponent in Eqn. (4.78) and remove the constraint by performing a summation over the appropriate variables (Navon et al., 2011).

$$Q = \sum_{N=0}^{\infty} \sum_{\{n_\alpha\}, \sum_\alpha n_\alpha = N} \exp(-\beta(\sum_\alpha n_\alpha E_\alpha - \mu N)) \quad (4.51)$$

The expression can be simplified to:

$$Q = \sum_{\{n_\alpha\}} \exp(\sum_\alpha n_\alpha \ln(1 - e^{\beta(\mu - E_\alpha)})). \quad (4.52)$$

Furthermore, it is worth noting the fact the occupation data serve as silent summation indices, allowing us to collectively represent them as (n). Therefore, the overall partition function (Q) may be expressed as:

$$Q = \prod_\alpha \sum_{n_\alpha} \exp(n_\alpha \ln(1 - e^{\beta(\mu - E_\alpha)})). \quad (4.53)$$

When dealing with fermions $((n = 0,1))$, the expression becomes $(1 + e^{\beta(\mu - E_\alpha)})$.. On the other hand, for bosons $((n = 0,1,2,3,\ldots))$, it takes the form of a series of angles with a sum of $((1 - e^{\beta(\mu - E_\alpha)})^{-1})$. By incorporating a parameter $(\theta = +1)$ over bosons and $(\theta = -1)$ for fermions, the expression for Q is given as:

$$Q = \prod_\alpha (1 - \theta e^{\beta(\mu - E_\alpha)})^{-\theta}. \quad (4.54)$$

The grand potential, which is equivalent to Eqn. (4.22), establishes a connection to the laws of thermodynamic (Navon et al., 2010):

$$J = PV = k_B T \ln Q = -\theta k_B T \sum_\alpha \ln(1 - \theta e^{\beta(\mu - E_\alpha)}) \quad (4.54)$$

This expression provides a comprehensive framework for determining the thermal characteristics of non-interacting bosons/fermions. In the realm of non-interacting permitted parts, the single-particle states (α) are distinguished from the single-particle momentum (p), which may include internal degrees of freedom. In the realm of low temperatures, the intricate internal dynamics tend to become stagnant, leaving behind just the ground state degeneracies that are identified through a spin variable (m_s). One example of such a variable is the (z)–projection corresponding to total angular momentum (Javanainen & Ruostekoski, 1995).

In reality, like Eqn. (4.40), the (α) sum involves a summation in (n_x, n_y, n_z), and potentially m_s. Given the dense nature of translational levels, one may substitute the sum over (n) with integration across energy (E), taking into account the weighting factor of the density for translational states $(g_{tr}(E))$,, as described in Eqn. (4.12).

$$\sum_\alpha \rightarrow \sum_{m_s} \sum_{n_x, n_y, n_z} \rightarrow g_s \int_0^\infty dE\ g_{tr}(E) \rightarrow \int_0^\infty dE\ g(E). \quad (4.56)$$

Here, the symbol $(g(E) = g_s g_{tr}(E))$ represents the overall density of states. Through this integration, a formula of state is derived:

$$P = \frac{J}{V} = -\theta k_B T \int_0^\infty dE\ g(E) \ln(1 - \theta e^{\beta(\mu - E)}).$$

Eqn. (4.91) provides a description of the pressure (P) in terms of temperature (T), volume (V), chemical potential (μ), and the characteristics of the single-particle states (E) and $(g(E))$. One way to express this integral is by using the dimensionless variables $(\tilde{E} = \beta E)$ and then applying integration by parts to solve it (Sotnikov et al., 2017).

$$P = -\theta(k_B T)^{5/2} \frac{g_s}{(2M)^{3/2} 4\pi^2} \int_0^\infty d\tilde{E}\, \tilde{E}\, \widetilde{E^{1/2}} \ln\left(1 - \theta e^{\beta(\mu - \tilde{E})}\right) \quad (4.57)$$

The equation presented here relates the pressure for an ideal gas to temperature (T), chemical potential (μ), and the thermal length (Λ), as stated in Eqn. (4.43). Nevertheless, determining (μ) experimentally poses significant challenges, which is why it is more advantageous to express \ P) in relation to (T) the density $([N]/V)$,and particularly in the high-temperature limit (Eqn. (4.47)). Regrettably, the absence of a straightforward analytical expression over (μ)in relation to density adds complexity to the task of deriving a convenient clear equation of state (Nascimbene et al., 2010).

It is worth noting that the computation of the internal power (U) confirms the elevated temperatures ideal gas relation:

$$P = \frac{2}{3}\frac{U}{V}. \quad (4.58)$$

This relationship remains valid at all temperatures for each boson and fermion, even though Eqn. (4.57) shows a deviation via the ideal behavior of gases at low temperatures. Upon investigation of the primary correction at high temperatures to the behavior of an ideal gas caused by quantum statistics, it is discovered (LeClair, 2008):

$$P = \frac{[N]}{V} k_B T \left[1 - \theta \frac{2^{5/2} - \theta}{\Lambda^3 [N]/V} + O\left(\frac{1}{\Lambda^6}\right)\right], \text{for } \Lambda^3 \frac{[N]}{gsV} = (2\pi)^{3/2} \frac{(M k_B T)^{3/2}}{[N]/V} \quad (4.59)$$

The explicit expression of the "degeneracy" parameters $(\Lambda^3 \frac{[N]}{gsV})$makes it clear that Eqn. (4.59) is an approximation applicable to high temperatures with low densities. The sign displayed by the leading correction suggests contrasting effects of different types of particle statistics on pressure. This behavior at high temperatures offers a glimpse into the contrasting characteristics of bosons or fermions, often referred to as there "social character." It can be quite difficult to experimentally confirm these adjustments in real gases, like atoms because the interactions between particles can introduce similar adjustments that are just as significant (Chávez et al., 2011).

When temperatures are low, the behaviors of ideal bosons along with fermions diverge significantly. The average work number $([n_\alpha])$ for single-particle states provides a clear indication of those distinctions (Yan & Blume, 2014). The value of this quantity is determined by:

$$[n_\alpha] = \frac{1}{e^{\beta(E_\alpha - \mu)} \mp 1}, \quad (4.60)$$

where; the ($\mp$) sign corresponds to bosons and fermions (Fermi-Dirac distribution, Eqn. (4.60)). The average occupation of each single-particle state depends solely on temperature and the energy of the state itself (De Rosi et al., 2017). However, the presence of other particles influences these distributions through the chemical potential (μ), which in turn depends on the total particle density ($[N]/V$) and temperature (T).

In which the ($\mp$) sign reflects particles with different properties (Fermi-Dirac distribution, Eqn. (4.60)). The distribution of particles among individual states is determined by the temperature along with the energy of each state (Algin & Arikan, 2017). Nevertheless, the distributions of particles are affected by the presence of extra particles due to the chemical potential (μ), which is determined by the overall particle density (N/V) and temperature (T).

ACTIVITY 4.1.

Explore the behavior of Fermi and Bose gases at low temperatures. Investigate how quantum statistics dictate the properties of these systems, such as Fermi-Dirac and Bose-Einstein distributions, which govern the behavior of fermions and bosons, respectively.

SUMMARY

- This chapter begins by introducing the basic principles of probability and statistics, emphasizing their application in describing the collective behavior of particles in systems.
- The chapter progresses to explore quantum statistics and the density operator, which are pivotal in analyzing the statistical mechanics of particles governed by quantum principles. It discusses equilibrium ensembles and their connection to thermodynamics, highlighting entropy as a key thermodynamic quantity that governs the direction of natural processes.
- Ideal systems are examined in detail, focusing on their behavior across different temperature regimes. The chapter distinguishes between high-temperature conditions, where classical statistical mechanics apply, and low-temperature scenarios dominated by quantum effects, particularly Fermi and Bose gases.
- Throughout the chapter, the concepts are illustrated with theoretical frameworks and mathematical formulations, providing a comprehensive foundation for further studies in statistical physics and thermodynamics.

REVIEW QUESTIONS

1. What are the fundamental concepts introduced in statistical physics?
2. How does quantum statistics differ from classical statistics?
3. Discuss the relationship between entropy and the second law of thermodynamics.
4. Explain the concept of equilibrium ensembles. How are they related to thermodynamic states?
5. Compare and contrast the behavior of Fermi gases and Bose gases at low temperatures.
6. What is the role of the density operator in quantum statistical mechanics?

MULTIPLE CHOICE QUESTIONS

1. **Which statistical concept is crucial for describing the behavior of particles in statistical physics?**
   a. Entropy
   b. Probability
   c. Density
   d. Energy
2. **At what temperature do Bose gases typically exhibit Bose-Einstein condensation?**
   a. Low temperatures
   b. High temperatures
   c. Room temperature
   d. Critical temperature
3. **What does the density operator represent in quantum statistical mechanics?**
   a. Probability distribution
   b. Energy eigenstates
   c. Density of particles
   d. State of the system
4. **Entropy is a measure of:**
   a. Disorder in a system
   b. Total energy of a system
   c. Temperature of a system
   d. Pressure of a system
5. **Equilibrium ensembles in statistical physics are associated with:**
   a. Constant energy systems
   b. Isolated systems
   c. Systems in thermal equilibrium
   d. Nonequilibrium systems

Answers to Multiple Questions

1. (b); 2. (a); 3. (d); 4. (a); 5. (c).

REFERENCES

1. Algin, A., & Arikan, A. S. (2017). Effective approach for taking into account interactions of quasiparticles from the low-temperature behavior of a deformed fermion-gas model. *Journal of Statistical Mechanics: Theory and Experiment, 2017*(4), 043105.
2. Algin, A., & Senay, M. (2012). High-temperature behavior of a deformed Fermi gas obeying interpolating statistics. *Physical Review E: Statistical, Nonlinear, and Soft Matter Physics, 85*(4), 041123.
3. Balcerzak, M., Dems, K., & Komisarski, A. (2007). Statistical convergence and ideal convergence for sequences of functions. *Journal of Mathematical Analysis and Applications, 328*(1), 715–729.
4. Blanciforti, L., & Green, R. (1983). An almost ideal demand system incorporating habits: An analysis of expenditures on food and aggregate commodity groups. *The Review of Economics and Statistics, 4*, 511–515.
5. Blatter, G., & Ivlev, B. I. (1994). Quantum statistical mechanics of vortices in high-temperature superconductors. *Physical Review B, 50*(14), 10272.
6. Buxton, R. B. (2021). The thermodynamics of thinking: Connections between neural activity, energy metabolism, and blood flow. *Philosophical Transactions of the Royal Society B, 376*(1815), 20190624.
7. Cahn, S. B., Mahan, G. D., & Nadgorny, B. E. (1997). *Thermodynamics and Statistical Physics* (Vol. 3, pp. 83–242). Springer US.
8. Castellano, C., Fortunato, S., & Loreto, V. (2009). Statistical physics of social dynamics. *Reviews of Modern Physics, 81*(2), 591–646.
9. Chapman, S. T., Fontana, M., Geroldinger, A., & Olberding, B. (2016). Multiplicative ideal theory and factorization theory. *Proceedings in Mathematics and Statistics, 170*, 65–69.
10. Chávez, I., Grether, M., & De Llano, M. (2011). Low-dimensional Fermi and Bose gases. *Physica E: Low-dimensional Systems and Nanostructures, 44*(2), 394–399.
11. Cimini, G., Squartini, T., Saracco, F., Garlaschelli, D., Gabrielli, A., & Caldarelli, G. (2019). The statistical physics of real-world networks. *Nature Reviews Physics, 1*(1), 58–71.
12. Cole, D. C. (2002). Connections between thermodynamics, statistical mechanics, quantum mechanics, and special astrophysical processes. *Fundamental Theories of Physics, 126*, 111–124.
13. Collell, G., & Fauquet, J. (2015). Brain activity and cognition: A connection from thermodynamics and information theory. *Frontiers in Psychology, 6*, 818.
14. Cowan, B. (2019). On the chemical potential of ideal Fermi and Bose gases. *Journal of Low Temperature Physics, 197*(5), 412–444.
15. Curado, E. M., & Tsallis, C. (1991). Generalized statistical mechanics: Connection with thermodynamics. *Journal of Physics A: Mathematical and General, 24*(2), L69.
16. Dacunha-Castelle, D., & Duflo, M. (2012). *Probability and Statistics: Volume II* (Vol. 2, pp. 54–60). Springer Science & Business Media.

17. De Ninno, G., & Fanelli, D. (2012). Out-of-equilibrium statistical ensemble inequivalence. *Europhysics Letters, 97*(2), 20002.

18. De Rosi, G., Astrakharchik, G. E., & Stringari, S. (2017). Thermodynamic behavior of a one-dimensional Bose gas at low temperature. *Physical Review A, 96*(1), 013613.

19. Dekking, F. M., Kraaikamp, C., Lopuhaä, H. P., & Meester, L. E. (2006). *A Modern Introduction to Probability and Statistics: Understanding Why and How* (Vol. 3, pp. 65–70). Springer Science & Business Media.

20. DeWitt, H. E. (1966). Statistical mechanics of high-temperature quantum plasmas beyond the ring approximation. *Journal of Mathematical Physics, 7*(4), 616–626.

21. Doob, J. L. (1934). Probability and statistics. *Transactions of the American Mathematical Society, 36*(4), 759–775.

22. Esposito, M., Harbola, U., & Mukamel, S. (2009). Nonequilibrium fluctuations, fluctuation theorems, and counting statistics in quantum systems. *Reviews of Modern Physics, 81*(4), 1665–1702.

23. Eu, B. C. (2013). *Nonequilibrium Statistical Mechanics: Ensemble Method* (Vol. 93, pp. 78). Springer Science & Business Media.

24. Fano, U. (1957). Description of states in quantum mechanics by density matrix and operator techniques. *Reviews of Modern Physics, 29*(1), 74.

25. He, P., & Kang, D. B. (2010). Entropy principle and complementary second law of thermodynamics for self-gravitating systems. *Monthly Notices of the Royal Astronomical Society, 406*(4), 2678–2688.

26. Hofmann, H. F. (2020). What does the operator algebra of quantum statistics tell us about the objective causes of observable effects? *Entropy, 22*(6), 638.

27. Ilievski, E., Quinn, E., & Caux, J. S. (2017). From interacting particles to equilibrium statistical ensembles. *Physical Review B, 95*(11), 115128.

28. Ishihara, M. (2021). Derivation of the density operator with quantum analysis for the generalized Gibbs ensemble in quantum statistics. *Physica A: Statistical Mechanics and its Applications, 583*, 126321.

29. Javanainen, J., & Ruostekoski, J. (1995). Off-resonance light scattering from low-temperature Bose and Fermi gases. *Physical Review A, 52*(4), 3033.

30. Johnson, R. A., Miller, I., & Freund, J. E. (2000). *Probability and statistics for engineers* (Vol. 2000, pp. 642). Pearson Education.

31. Kapur, J. N., & Kesavan, H. K. (1992). Entropy optimization principles and their applications. In *Entropy and Energy Dissipation in Water Resources* (Vol. 3, pp. 3–20). Dordrecht: Springer Netherlands.

32. Krekelberg, W. P., Siderius, D. W., Shen, V. K., Truskett, T. M., & Errington, J. R. (2013). Connection between thermodynamics and dynamics of simple fluids in highly attractive pores. *Langmuir, 29*(47), 14527–14535.

33. LeClair, A. (2008). Interacting Bose and Fermi gases in low dimensions and the Riemann hypothesis. *International Journal of Modern Physics A, 23*(09), 1371–1391.

34. Lee, P. A. (2007). From high temperature superconductivity to quantum spin liquid: Progress in strong correlation physics. *Reports on Progress in Physics, 71*(1), 012501.
35. Leff, H. S. (1969). Entropy differences between ideal and nonideal systems. *American Journal of Physics, 37*(5), 548–553.
36. Luzzi, R., Vasconcellos, Á. R., & Ramos, J. G. (2002). Predictive statistical mechanics: A nonequilibrium ensemble formalism. *Springer Science & Business Media* (Vol. 122, pp. 321–325).
37. Lyubartsev, A. P., & Laaksonen, A. (2004). On the reduction of molecular degrees of freedom in computer simulations. In *Novel Methods in Soft Matter Simulations* (Vol. 3, pp. 219–244). Berlin, Heidelberg: Springer Berlin Heidelberg.
38. Man'ko, V. I., Man'ko, O. V., & Chernega, V. N. (2020). Conventional quantum statistics with a probability distribution describing quantum system states. *Physics of Particles and Nuclei, 51*, 772–780.
39. Martyushev, L. M., & Seleznev, V. D. (2006). Maximum entropy production principle in physics, chemistry and biology. *Physics Reports, 426*(1), 1–45.
40. Mézard, M., & Parisi, G. (2000). Statistical physics of structural glasses. *Journal of Physics: Condensed Matter, 12*(29), 6655.
41. Miller, J., & Miller, J. (2019). Probability and statistics. *Planetary Spacecraft Navigation, 6*, 187–215.
42. Minlos, R. A. F. (1968). Lectures on statistical physics. *Russian Mathematical Surveys, 23*(1), 137.
43. Nascimbene, S., Navon, N., Chevy, F., & Salomon, C. (2010). The equation of state of ultracold Bose and Fermi gases: A few examples. *New Journal of Physics, 12*(10), 103026.
44. Navon, N., Nascimbene, S., Chevy, F., & Salomon, C. (2010). The equation of state of a low-temperature Fermi gas with tunable interactions. *Science, 328*(5979), 729–732.
45. Navon, N., Piatecki, S., Günter, K., Rem, B., Nguyen, T. C., Chevy, F., & Salomon, C. (2011). Dynamics and thermodynamics of the low-temperature strongly interacting Bose gas. *Physical Review Letters, 107*(13), 135301.
46. Ozawa, H., Ohmura, A., Lorenz, R. D., & Pujol, T. (2003). The second law of thermodynamics and the global climate system: A review of the maximum entropy production principle. *Reviews of Geophysics, 41*(4).
47. Pathria, R. K. (1998). Similarities and differences between Bose and Fermi gases. *Physical Review E, 57*(3), 2697.
48. Pérez, E., & Sauer, T. (2010). Einstein's quantum theory of the monatomic ideal gas: Non-statistical arguments for a new statistic. *Archive for History of Exact Sciences, 64*, 561–612.
49. Pople, J. A. (1954). The statistical mechanics of assemblies of axially symmetric molecules-I. General theory. *Proceedings of the Royal Society of London. Series A. Mathematical and Physical Sciences, 221*(1147), 498–507.
50. Rosenkrantz, R. D. (2012). *ET Jaynes: Papers on Probability, Statistics and Statistical Physics*

(Vol. 158, pp. 89–90). Springer Science & Business Media.

51. Ross, S. (2009). *Probability and Statistics for Engineers and Scientists* (16th ed., pp. 32–33). Elsevier.

52. Sands, D., & Dunning-Davies, J. (2014). Information entropy and the statistical physics of the classical ideal gas. *Physics Essays, 27*(4), 537–547.

53. Schirmer, S. G. (2000). Laser cooling of internal molecular degrees of freedom for vibrationally hot molecules. *Physical Review A, 63*(1), 013407.

54. Schwabl, F. (2006). Equilibrium ensembles. In *Statistical Mechanics* (Vol. 2, pp. 25–74). Berlin, Heidelberg: Springer Berlin Heidelberg.

55. Shah, J. K., & Maginn, E. J. (2011). A general and efficient Monte Carlo method for sampling intramolecular degrees of freedom of branched and cyclic molecules. *The Journal of Chemical Physics, 135*(13).

56. Sommerfeld, A. (2012). *Lectures on Theoretical Physics: Thermodynamics and Statistical Mechanics* (Vol. 5, pp. 21–25). Academic Press.

57. Sotnikov, A. G., Sereda, K. V., & Slyusarenko, Y. V. (2017). Chemical potentials and thermodynamic characteristics of ideal Bose-and Fermi-gases in the region of quantum degeneracy. *Low Temperature Physics, 43*(1), 144–151.

58. Stratt, R. M. (1984). The statistical mechanics of a liquid of two-state molecules. *The Journal of Chemical Physics, 80*(11), 5764–5775.

59. Suhov, Y. M., & Kelbert, M. (2014). *Probability and Statistics by Example* (Vol. 1, pp. 12–15). Cambridge University Press.

60. Tanaka, F., & Komaki, F. (2005). Bayesian predictive density operators for exchangeable quantum-statistical models. *Physical Review A—Atomic, Molecular, and Optical Physics, 71*(5), 052323.

61. Wang, L. S. (2017). The second law: From Carnot to Thomson-Clausius, to the theory of exergy, and to the entropy-growth potential principle. *Entropy, 19*(2), 57.

62. Wang, Z., Bauch, C. T., Bhattacharyya, S., d'Onofrio, A., Manfredi, P., Perc, M., & Zhao, D. (2016). Statistical physics of vaccination. *Physics Reports, 664*, 1–113.

63. Yan, Y., & Blume, D. (2014). Temperature dependence of small harmonically trapped atom systems with Bose, Fermi, and Boltzmann statistics. *Physical Review A, 90*(1), 013620.

64. Zurek, W. H. (2018). Eliminating ensembles from equilibrium statistical physics: Maxwell's demon, Szilard's engine, and thermodynamics via entanglement. *Physics Reports,* 755, 1–21.

CHAPTER 5

Solids

LEARNING OBJECTIVES

After studying this chapter, you will be able to:

1. Understand the microscopic structure of solids, including lattices and crystal structures;
2. Explore the concept and significance of the reciprocal lattice in solid-state physics;
3. Analyze the principles and techniques of diffraction experiments used to study crystal structures;
4. Investigate the behavior of electrons in crystals and models of electronic bands;
5. Examine the filling of electronic bands and the distinction between metals and insulators;

6.3Study the vibrations of crystals, including normal modes, thermal properties, and other phonon effects.

KEY TERMS FROM THIS CHAPTER

Crystal structure
Electronic bands
Normal modes
Reciprocal lattice
Diffraction
Lattice
Phonons

UNIT INTRODUCTION

Solids, a fundamental focus in material science, include a wide range of materials from crystalline lattices with precise periodicity to amorphous structures without long-range order. This chapter delves into the atomic structures that influence their macroscopic behaviors and technological applications (Bresler & Dagan, 1983).

The first section of the chapter focuses on crystalline solids, which are defined by their orderly atomic arrangements within distinct lattice structures. Topics covered include crystal systems, symmetry operations, and the influence of defects and impurities on material properties such as mechanical strength, electrical conductivity, and thermal expansion. The systematic analysis of crystallographic principles and their application in various materials provides a solid foundation for understanding how atomic-scale arrangements determine the observed properties and behavior of crystalline solids under external conditions (Smith & Oatley, 1955).

The chapter then explores amorphous solids, where atoms lack long-range periodicity. This section examines the structural characteristics of glasses and polymers, focusing on the absence of crystalline order and its impact on material properties. Discussions include the formation mechanisms of amorphous solids, their structural stability, and the relationship between local atomic configurations and macroscopic behaviors. By comparing crystalline and amorphous structures, the chapter highlights the significant role of atomic arrangement in determining material functionality, emphasizing their importance in technological applications such as semiconductors and biomedical materials (Apweiler et al., 2009).

Remember

Understanding the foundational concepts of neurons, neural networks, and their training algorithms like backpropagation is crucial for effectively implementing and optimizing deep learning models in various applications.

5.1. THE MICROSCOPIC STRUCTURE OF SOLIDS

The microscopic structure of solids forms the foundation of their macroscopic properties and behaviors. At this scale, solids reveal arrangements of atoms or molecules tightly bound together through chemical bonds. In crystalline solids, these arrangements form well-defined lattice structures with periodicity, imparting materials with distinct mechanical, electrical, and thermal properties. Conversely, amorphous solids lack such long-range order, exhibiting random atomic configurations that contribute to unique properties like high optical transparency or mechanical flexibility (Stokes & Hatch, 2002) (Figure 5.1).

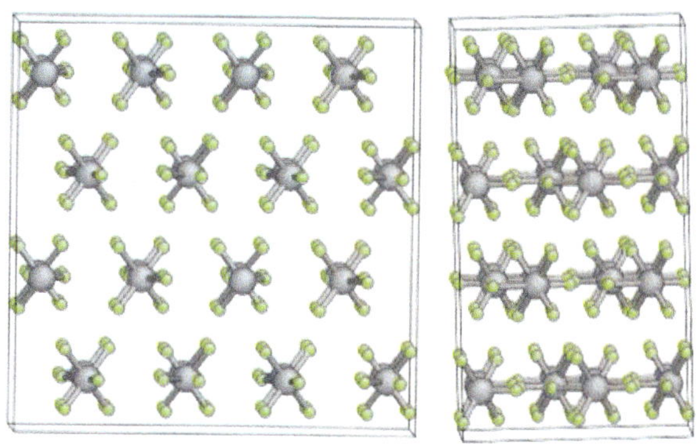

Figure 5.1. *The crystal structure of UF6.*

Source: B. Morel, Creative Commons License.

5.1.1. Lattices and Crystal Structures

In the study of solids, particularly crystalline materials, the concept of lattices and crystal structures forms the basis of understanding their physical properties and behaviors. This section explains the fundamental principles governing the spatial arrangement of atoms or ions within crystalline solids, explaining the mathematical frameworks and structural classifications that define their unique characteristics (Paukstelis et al., 2004).

5.1.1.1. Mathematical Formulation of Lattices

A lattice in the context of crystallography refers to an infinite array of points in space that represent the positions of atoms, ions, or molecules within a crystal. Mathematically, a lattice can be defined as:

$$R = n_1a_1 + n_2a_2 + n_3a_3$$

where; (R) is the position vector within the lattice; (n_1, n_2, n_3) are integers; and (a_1, a_2, a_3) are basis vectors spanning the lattice. These basis vectors define the unit cell, the smallest repeating unit that, when translated by lattice vectors, recreates the entire crystal structure.

The lattice vectors (a_1, a_2, a_3) can be chosen arbitrarily, but conventionally, they are selected to maintain symmetry and periodicity in the crystal lattice. The lattice points are positioned at regular intervals defined by these vectors, ensuring that the crystal lattice exhibits translational symmetry in all directions (Epstein, 1960).

5.1.1.2. Types of Lattices

Crystalline materials can be categorized into several types of lattices based on their symmetry and atomic arrangements. The most common types include (Figure 5.2):

1. **Simple Cubic Lattice (SC):** In a simple cubic lattice, each lattice point is surrounded by eight nearest neighbors, situated at the corners of a cube. The lattice parameters (a) represent the distance between adjacent lattice points along each axis.

$$R = n_1 a + n_2 b + n_3 c$$

where; (a, b, c) are the lattice vectors; and (n_1, n_2, n_3) are integers.

2. **Body-Centered Cubic Lattice (BCC):** The body-centered cubic lattice adds an additional lattice point at the center of the cube, besides the points at the cube corners. This arrangement results in a coordination number of 8, where each lattice point is surrounded by 8 nearest neighbors (Rota, 1997).

$$R = n_1 a + n_2 b + n_3 c + \frac{1}{2}(a + b + c)$$

3. **Face-Centered Cubic Lattice (FCC):** The face-centered cubic lattice places additional lattice points at the centers of each face of the cube, in addition to those at the corners. This results in a coordination number of 12, with each lattice point surrounded by 12 nearest neighbors.

$$R = n_1 a + n_2 b + n_3 c + \frac{1}{2}(a + b), \frac{1}{2}(b + c), \frac{1}{2}(c + a)$$

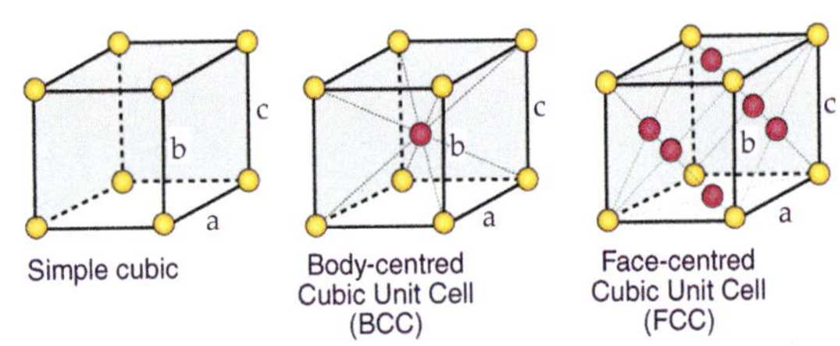

Figure 5.2. *Types of lattices.*

Source: BYJU's, Creative Commons License.

5.1.1.3. Crystal Structures

Crystal structures refer to the specific arrangement of atoms, ions, or molecules within a crystal lattice.

These structures are defined by the positions of lattice points and the types of atoms or ions occupying them. The classification of crystal structures is based on the symmetry operations that leave the crystal unchanged, including translations, rotations, and reflections (Rota, 1997).

1. **Primitive Cell:** The simplest repeating unit within a crystal lattice, containing only one lattice point.

$$V_{\text{cell}} = |a_1 \cdot (a_2 \times a_3)|$$

where; (a_1, a_2, a_3) are the lattice vectors defining the primitive cell; and (V_{cell}) is the volume of the primitive cell.

2. **Unit Cell:** The smallest repeating unit in a crystal lattice, encompassing a specific number of lattice point (Figure 5.3).

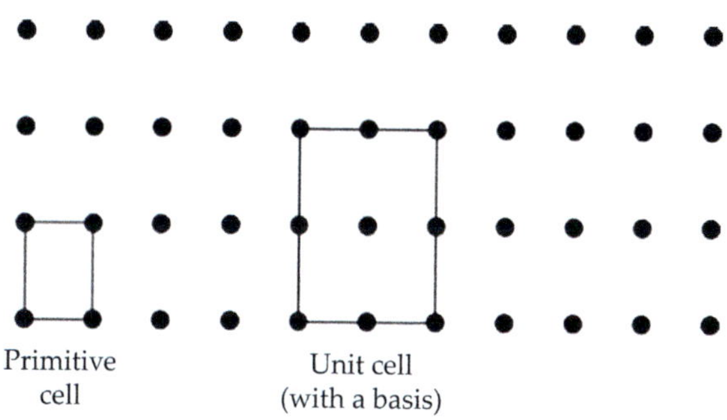

Figure 5.3. *Difference between a primitive cell and a unit cell for a rectangular lattice.*

Source: Prasanta K, Creative Commons License.

5.1.2. The Reciprocal Lattice

In the study of crystalline solids, the reciprocal lattice plays a pivotal role in understanding the physical properties and behaviors of materials, particularly in the context of diffraction phenomena and electronic band structures. This section explores the mathematical formulation and significance of the reciprocal lattice, explaining its relationship with the real-space lattice and its utility in describing wave interactions within crystals (Ebalard & Spaepen, 1989).

5.1.2.1. Mathematical Formulation

The reciprocal lattice of a crystal is defined as the set of all wave vectors (G) that satisfy the condition:

$$G \cdot R = 2\pi n$$

where; (R) is a vector in the real-space lattice; (G) is a vector in the reciprocal lattice; and (n) is an integer. This condition ensures that (G) is orthogonal to all vectors (R) in the real-space lattice (Pasinetti, 1960).

The reciprocal lattice vectors (G) are related to the real-space lattice vectors (a_i) (where $(i = 1, 2, 3)$) through the following relations:

$$b_i = 2\pi \frac{a_j \times a_k}{a_i \cdot (a_j \times a_k)}$$

where; (b_i) are the reciprocal lattice vectors corresponding to (a_i); and (a_j) and (a_k) are the other two lattice vectors not equal to (a_i).

5.1.2.2. Properties and Applications

The reciprocal lattice provides a powerful framework for understanding diffraction patterns observed in crystals, such as those in X-ray diffraction experiments. According to Bragg's law, constructive interference of X-rays occurs when the condition $(k \cdot G = 2\pi)$ is met, where (k) is the wave vector of the incident X-ray beam and (G) is a reciprocal lattice vector. This relationship allows researchers to determine the spacing between atomic planes in a crystal, known as the d-spacing, from the positions of diffraction peaks (Datta & Grant, 2004).

5.1.2.3. Relation to Real-Space Properties

The reciprocal lattice vectors (G) provide valuable insights into the physical properties of crystals in real space. For instance, the density of states and phonon dispersion relations can be calculated using information derived from the reciprocal lattice structure. The phonon dispersion relation, which describes the energy dependence of vibrational modes in a crystal, is directly related to the frequencies of phonons and their spatial periodicity in the reciprocal lattice.

Moreover, the reciprocal lattice facilitates the analysis of physical phenomena such as band folding, where multiple Brillouin zones overlap due to finite crystal sizes or specific boundary conditions. Band folding affects the electronic band structure and can lead to observable changes in transport properties like conductivity and thermal conductivity (Pendás et al., 2007).

The reciprocal lattice is a fundamental concept in the study of crystalline solids, offering a mathematical framework that connects the spatial arrangement of atoms in real space with wave interactions in reciprocal space. Its applications extend across various fields of materials science and physics, from diffraction analysis to electronic structure calculations, underscoring its critical role in understanding and manipulating the properties of materials at the atomic scale (Diamond, 1971).

5.1.3. Diffraction Experiments

Diffraction experiments are pivotal in elucidating the structural characteristics and atomic arrangements within crystalline solids. This section delves into the theoretical underpinnings of diffraction phenomena, detailing the mathematical frameworks and experimental methodologies employed to analyze X-ray, neutron, and electron diffraction patterns (Ungár, 2001).

Did you know?

Did you know that diffraction experiments, such as X-ray diffraction, are crucial for determining the atomic arrangement within crystals? These experiments use the wave nature of X-rays to probe the periodic structure of solids.

5.1.3.1. Bragg's Law

Bragg's law forms the foundation of diffraction analysis, describing the conditions under which X-rays or other coherent radiation exhibit constructive interference from a crystal lattice (Figure 5.4). According to Bragg's law, the condition for constructive interference is given by:

$$2d \sin\theta = n\lambda$$

where; (d) is the spacing between atomic planes in the crystal; $(\grave{e})$ is the angle of incidence of the X-ray beam relative to the crystal lattice planes; $(n)i$ s an integer representing the order of the diffraction peak; and $(\ddot{e})$ is the wavelength of the incident radiation (Weigel, 1983).

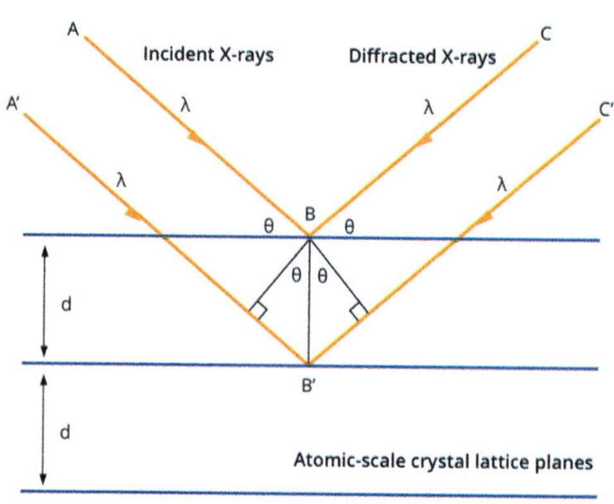

Figure 5.4. *Bragg's Law reflection. The diffracted X-rays exhibit constructive interference when the distance between paths ABC and A'B'C' differs by an integer number of wavelengths* (λ).

Source: Darrell Henry, Creative Commons License.

This relationship arises from the constructive interference of waves reflected by successive atomic planes within the crystal lattice. By measuring the angles (θ) at which diffraction peaks occur for known (λ) researchers can determine the spacing (d) between lattice planes, providing crucial structural information about the crystal (Shvyd'Ko & Lindberg, 2012).

5.1.3.2. Experimental Techniques

1. **X-Ray Diffraction:** X-ray diffraction is widely used to study the atomic structure of crystalline solids due to the high penetration depth of X-rays and their ability to interact with electron clouds in atoms. The diffraction pattern observed from a crystal provides information about the symmetry and periodic arrangement of atoms within the unit cell. The intensity and position of diffraction peaks are analyzed to deduce crystallographic parameters such as lattice constants, atomic positions, and crystal orientation (Lucas et al., 2018).
2. **Neutron Diffraction:** Neutron diffraction experiments utilize neutron beams to probe the atomic arrangement within materials. Neutrons interact with atomic nuclei through nuclear forces, making them sensitive to light elements such as hydrogen and providing complementary information to X-ray diffraction. Neutron diffraction is particularly valuable for studying magnetic structures and hydrogen bonding in materials, where traditional X-ray techniques may be less effective (Field et al., 2004).
3. **Electron Diffraction:** Electron diffraction involves directing a beam of electrons onto a crystalline sample and analyzing the resulting diffraction pattern. Due to the

short wavelength of electrons, electron diffraction offers extremely high spatial resolution, enabling detailed studies of atomic arrangements and defects within crystals. Transmission electron microscopy (TEM) combines electron diffraction with imaging capabilities, allowing direct visualization of crystal defects, dislocations, and nanoscale structures (Hotelling, 1944).

5.1.3.3. Practical Applications

Diffraction experiments have diverse applications across materials science, chemistry, and physics, contributing to advancements in various fields:

1. **Materials Characterization:** Diffraction techniques are crucial for determining the crystal structure and phase purity of materials, essential for the development of new materials with tailored properties.
2. **Defect Analysis:** By analyzing deviations from ideal crystal structures in diffraction patterns, researchers can identify and characterize defects such as vacancies, dislocations, and stacking faults within crystals.
3. **Phase Identification:** Diffraction patterns serve as fingerprints for different crystalline phases, aiding in the identification and quantification of phases in complex materials like alloys and composite materials (Båth, 2010).

In conclusion, diffraction experiments provide invaluable insights into the atomic and molecular arrangements within crystalline solids, enabling precise characterization of their structural, electronic, and magnetic properties. By leveraging mathematical principles and advanced experimental techniques, researchers continue to push the boundaries of materials science, uncovering new phenomena and facilitating the development of novel materials for diverse applications.

5.2. ELECTRONS IN CRYSTAL

In crystalline solids, electrons occupy distinct energy states governed by the periodic potential of the crystal lattice. These energy states form electronic bands, where electrons exhibit wave-like behavior and are delocalized over the entire crystal. The interaction between electrons and the periodic potential of the lattice gives rise to band structures, which dictate the electrical conductivity, optical properties, and electronic behavior of the material. Understanding the behavior of electrons in crystals is crucial for developing semiconductor devices, understanding superconductivity, and exploring novel electronic phenomena such as quantum confinement effects in nanomaterials (Davisson, 1928).

5.2.1. Models of Bands in Crystals

The modeling of electronic band structures in crystals is essential for understanding the complex behaviors of electrons within solid-state materials. This section explores various theoretical frameworks and models used to describe the formation and characteristics of electronic bands in crystalline solids, highlighting their mathematical formulations and implications for material propertie (Kosobutskii & Poplavnoi, 2006).

5.2.1.1. Nearly Free Electron Model

The nearly free electron (NFE) model is a simplified approach that describes the behavior of electrons in a periodic potential of a crystal lattice. In this model, electrons are treated as free particles with slight perturbations due to the periodic potential of the lattice. The potential energy $V(r)$ experienced by an electron at position (r) in the crystal is periodic:

$$V(r) = V(r + R)$$

where; (R) is a lattice vector. The Schrödinger equation for an electron in this potential can be approximated as:

$$\left[-\frac{\hbar^2}{2m}\nabla^2 + V(r)\right]\psi(r) = E\psi(r)$$

The Bloch theorem states that the solutions to this equation can be expressed as:

$$\psi_k(r) = e^{ik\cdot r}u_k(r)$$

where; $\psi_k(r)$ is the Bloch wave function, (k) is the wave vector in reciprocal space, and $(u_k(r))$ is a periodic function with the same periodicity as the lattice (Khazaei et al., 2016).

The energy dispersion relation $E(k)$, or the band structure, in the NFE model is approximated by:

$$E(k) = \frac{\hbar^2 |k|^2}{2m} + E_0$$

where; (E_0) is the energy offset due to the periodic potential.

5.2.1.2. Tight Binding Model

The Tight Binding (TB) model provides a more detailed description of electronic bands by considering the localized nature of atomic orbitals and their overlap in forming molecular-like orbitals within a crystal. In this model, electrons are treated as being primarily localized near atomic sites and interact through hopping integrals (t_{ij}) between neighboring atoms (i) and (j).

The Hamiltonian for the tight binding model is:

$$H = \sum_i \epsilon_i c_i^{\dagger} c_i + \sum_{\langle i,j \rangle} t_{ij} c_i^{\dagger}$$

where; (ϵ_i) is the energy of the atomic orbital at site (i),$(c_i^{\dagger})$ and (c_i) are the creation and annihilation operators for electrons at site (i), and $\langle i,j \rangle$ denotes summation over nearest neighbor pairs.

The energy eigenvalues $(E_{k,n})$ and corresponding wave functions $(\psi_{k,n}(r))$ in the tight binding model are obtained by solving the Schrödinger equation (Ridolfi et al., 2015):

$$\sum_j t_{ij} e^{ik\cdot(R_i - R_j)} \psi_{k,n}(r) = E_{k,n} \psi_{k,n}(r)$$

where; (R_i) and (R_j) are the positions of atoms (i) and (j) in the crystal lattice.

5.2.1.3. Other Models and Extensions

Beyond the NFE and TB models, numerous extensions and modifications exist to describe specific properties of electronic bands in crystals. These include:

1. **Kronig-Penney Model:** A periodic potential step function that approximates the crystal potential.
2. **Effective Mass Approximation:** Describes electrons in semiconductors and metals as free particles with effective masses.
3. **Density Functional Theory (DFT):** Uses the electron density to calculate the electronic band structure and other properties of materials (Pina et al., 2006).

These models and approaches provide varying levels of complexity and accuracy in describing the electronic properties of solids, each tailored to different material systems and phenomena. By understanding and manipulating these band structures, researchers can predict and engineer the electronic, optical, and magnetic properties of materials for applications in semiconductor devices, photonics, and quantum technologies.

5.2.2. Filling of the Bands: Metals and Insulators

The filling of electronic bands in crystalline solids fundamentally determines their electrical conductivity and classification as metals, insulators, or semiconductors. This section explores the principles governing band filling in metals and insulators, emphasizing the role of Fermi-Dirac statistics, band overlap, and band gaps in defining material properties (Issendorff & Cheshnovsky, 2005).

Tip

In materials, the filling of electronic energy bands determines whether a substance behaves as a metal (conducting) or an insulator (non-conducting). Understanding band theory helps predict and explain the electrical properties of solids.

5.2.2.1. Metals: Fermi Level and Conductivity

In metals, the electronic bands are partially filled with electrons that are free to move throughout the crystal lattice, contributing to high electrical conductivity. The Fermi level (E_F) (Figure 5.5) represents the energy level at which the probability of finding an electron is $\left(\frac{1}{2}\right)$. According to Fermi-Dirac statistics, the distribution of electrons in bands follows:

$$f(E) = \frac{1}{1 + e^{(E-E_F)/k_B T}}$$

where; $f(E)$ is the Fermi-Dirac distribution function; (E) is the energy of the electronic state; $(k_B)$is the Boltzmann constant; and (T) is the temperature. At absolute zero temperature $(T=0)K)$, all states up to the Fermi level (E_F) are occupied, while states above (E_F) are empty. The density of states $(D(E))$ determines the number of energy states available per unit energy interval (Kurbatsky et al., 2009).

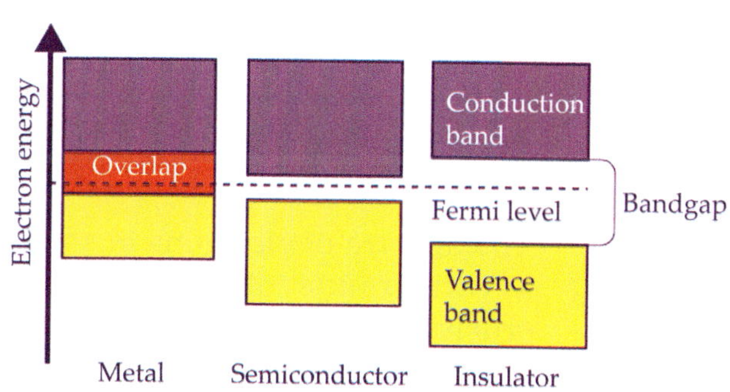

Figure 5.5. *Fermi level in metals, semiconductors, and insulators.*

Source: BYJU's, Creative Commons License.

The electrical conductivity (σ) in metals is high due to the presence of free electrons that can move through the crystal lattice without significant resistance. The conductivity can be related to the relaxation time (τ) of electrons and the effective mass (m^*) through the Drude model:

$$\sigma = \frac{ne^2\tau}{m^*}$$

where; (n) is the electron density; (e) is the elementary charge; and (m^*) is the effective mass of the electrons.

5.2.2.2. Insulators: Band Gaps and Electrical Insulation

In contrast to metals, insulators have filled valence bands and a large energy gap (E_g) separating the valence band from the conduction band. Electrons in insulators cannot move freely between bands due to the absence of available energy states within the band gap. The band gap (E_g) is typically on the order of electron volts (eV) and determines the electrical resistivity of the material.

The probability of finding electrons in the conduction band at temperature (T) can be described by:

$$n_c = N_c e^{-E_g/(k_B T)}$$

where; (N_c) is the effective density of states in the conduction band. At temperatures well below the band gap (E_g), the conduction band remains mostly empty, leading to very low electrical conductivity.

Insulators can be further classified into:

1. **Direct Band Gap Insulators:** The minimum energy of the conduction band is directly opposite the maximum energy of the valence band.
2. **Indirect Band Gap Insulators:** The minimum energy of the conduction band is not directly opposite the maximum energy of the valence band (Drummond et al., 2012).

5.2.2.3. Practical Implications and Applications

Metals find widespread use in electrical wiring, conductors, and electronic components where high conductivity is essential. Insulators are critical for electrical insulation, dielectric materials, and semiconductor devices where precise control over electron transport properties is necessary.

Band structure calculations through methods like Density Functional Theory (DFT) enable precise predictions of electronic properties based on the atomic structure and composition of materials. This computational approach aids in the development of new materials with optimized band structures for advanced technological applications, ranging from energy storage devices to optoelectronic devices (Kehr, 1993).

5.2.3. Spectra of Electrons in Solids

The spectra of electrons in solids encompass a wide range of phenomena that reveal crucial insights into the electronic structure, optical properties, and excitations within crystalline materials. This section explores the diverse spectra observed in solids, including optical absorption spectra, photoemission spectra, and phonon spectra, highlighting their theoretical foundations and experimental implications (Daniels et al., 1970).

5.2.3.1. Optical Absorption Spectra

Optical absorption spectra in solids provide information about the energy levels and electronic transitions within the material. When photons of sufficient energy are incident on a solid, they can excite electrons from the valence band to the conduction band, creating electron-hole pairs. The absorption coefficient $\alpha(\omega)$, which describes the material's ability to absorb light at a given frequency (ω), is related to the imaginary part of the dielectric function $(\epsilon(\omega))$:

$$\alpha(\omega) = \frac{2\omega}{c}\sqrt{\frac{\epsilon_2(\omega)}{2}}$$

where; $(\epsilon_2(\omega))$ is the imaginary part of the dielectric function; and (c) is the speed of light.

The optical absorption spectra provide insights into the band structure, band gaps, and electronic transitions within the material. Sharp absorption peaks correspond to direct transitions between energy levels, while broad absorption features may indicate indirect transitions involving phonons or defects (Stutz & Platt, 1996).

5.2.3.2. Photoemission Spectra

Photoemission spectroscopy (PES) involves the emission of electrons from a material's surface when illuminated with photons of sufficient energy. PES provides direct information about the electronic structure near the surface of the material, including the density of states (DOS) and the Fermi level (E_F). The kinetic energy (E_{kin}) of emitted electrons is related to the incident photon energy $(h\nu)$ and the work function (ϕ) of the material:

$$E_{\text{kin}} = h\nu - \phi$$

Analysis of photoemission spectra allows researchers to determine the binding energies of electrons and infer the band structure and electronic interactions within the material (Littlewood et al., 2010).

5.2.3.3. Phonon Spectra

Phonon spectra characterize the vibrational modes and lattice dynamics of solids, providing insights into thermal conductivity, thermal expansion, and lattice stability. Inelastic neutron scattering and Raman spectroscopy are commonly used experimental techniques to study phonon spectra.

The energy $(\hbar\omega)$ of a phonon mode is related to its wave vector (q) and the force constants (Φ_q) describing atomic interactions:

$$\hbar\omega_q = \sqrt{\frac{\Phi_q}{M}}$$

where; (M) is the mass of the atoms involved in the phonon mode.

Phonon spectra exhibit dispersion relations that describe the relationship between phonon energy and wave vector, revealing information about phonon lifetimes, anharmonic effects, and lattice thermal conductivity (Balandin, 2005).

5.2.3.4. Practical Applications and Advances

The study of electron spectra in solids underpins advancements in materials science and technology. Optical absorption spectra guide the development of photovoltaic devices and optoelectronic materials by optimizing light absorption and charge carrier generation. Photoemission spectra aid in the design of semiconductor devices and quantum materials by revealing electronic band structures and surface states. Phonon spectra contribute to the development of thermal management materials and phononic devices by controlling heat transport and thermal properties (Field, 1997).

Recent advances in spectroscopic techniques, such as ultrafast laser spectroscopy and scanning tunneling microscopy combined with spectroscopy, enable real-time studies of electronic and vibrational dynamics at the nanoscale. Computational methods, including Density Functional Theory (DFT) and ab initio calculations, complement experimental spectra by predicting and interpreting complex electronic and vibrational properties of materials.

In conclusion, the spectra of electrons in solids encompass a diverse range of experimental techniques and theoretical frameworks that elucidate fundamental electronic and vibrational properties. By integrating spectroscopic data with theoretical models, researchers continue to uncover new phenomena and engineer materials with tailored functionalities for applications in electronics, photonics, energy conversion, and beyond (Byrne et al., 2018).

5.3. THE VIBRATIONS OF CRYSTALS

The vibrations of crystals, known as phonons, are fundamental to understanding the thermal, mechanical, and electronic properties of solids. Phonons represent quantized lattice vibrations that propagate through the crystal structure, influencing phenomena such as thermal conductivity and heat capacity. These vibrational modes are characterized by their frequencies and wave vectors, providing insights into the material's stability and response to external stimuli. Studying phonons through techniques like neutron scattering and Raman spectroscopy allows researchers to explore lattice dynamics, phonon dispersion relations, and the coupling between vibrational modes and other physical properties in crystals (Cahill & Pohl, 1988).

5.3.1. The Normal Modes of Vibration

The normal modes of vibration in crystals, also known as phonons, are essential for understanding the dynamic behavior of solids at the atomic scale. This section delves into the theoretical foundations, mathematical descriptions, and physical implications of phonons, exploring their role in defining thermal conductivity, lattice stability, and the vibrational properties of crystalline materials (Luck & Petritis, 1986).

5.3.1.1. Phonons

Phonons are quantized collective excitations of atoms in a crystalline lattice, analogous to photons in the electromagnetic field. They arise from the periodic potential of the lattice and describe the vibrational modes that propagate through the solid. The dynamics of phonons are governed by Newton's second law, where the force (F_i) on an atom (i) is related to its displacement (u_i):

$$[M_i \frac{d^2 u_i}{dt^2} = -\sum_j \Phi_{ij} u_j]$$

where; (M_i) is the mass of atom (i); (Φ_{ij}) represents the force constant between atoms (i) and (j); and (u_i) is the displacement of atom (i) (Zeiger et al., 1992).

5.3.1.2. Mathematical Formulation

The vibrational modes of a crystal can be described using the concept of normal modes, where each mode corresponds to a specific pattern of atomic displacements.

The displacement of atom (i) in the (i)-th normal mode can be expressed as:

$$[u_i^\nu(t) = A_i^\nu e^{i(q \cdot r_i - \omega^\nu t)}]$$

where; (A_i^ν) is the amplitude of the mode (ν); (q) is the wave vector; (r_i) is the position vector of atom (i); (ω^ν) is the angular frequency of the mode; and (t) is time.

The angular frequency (ω^ν) of each normal mode is determined by the eigenvalue equation:

$$\sum_j \Phi_{ij} e^{iq \cdot (r_i - r_j)} = \omega^\nu M_i e^{iq \cdot r_i}$$

This equation encapsulates the vibrational dynamics of the crystal, where the forces between atoms (Φ_{ij}) lead to characteristic frequencies (ω^ν) associated with each normal mode (Pasinetti, 1960).

5.3.1.3. Brillouin Zone and Dispersion Relations

The Brillouin zone plays a crucial role in describing the propagation of phonons through reciprocal space. The dispersion relation $(\omega(q))$ relates the angular frequency (ω) of phonons to the wave vector (q):

$$[\omega(q) = \sqrt{\frac{\Phi_q}{M}}]$$

where; (Φ_q) is the force constant matrix in reciprocal space; and ( ) is the mass of the atoms involved in the phonon mode. The Brillouin zone boundary defines the maximum wave vector (q) for phonon propagation, influencing the phonon dispersion and contributing to the material's thermal and mechanical properties (Sivarajah et al., 2016).

5.3.1.4. Physical Implications and Applications

Understanding the normal modes of vibration is crucial for interpreting various physical properties of solids. Phonons determine the thermal conductivity and heat capacity of materials, as their propagation through the crystal lattice influences how thermal energy is transported. Additionally, phonons contribute to the elastic properties and mechanical stability of crystals, affecting their response to mechanical stress and strain (Li et al., 2001).

Experimental techniques such as inelastic neutron scattering and Raman spectroscopy enable direct observation and characterization of phonon dispersion relations and vibrational modes in crystals. Computational methods, including Density Functional Theory (DFT) and molecular dynamics simulations, complement experimental data by predicting phonon spectra and thermal properties based on atomic structure and interatomic interactions. Innovations in phonon engineering have led to advancements in thermoelectric materials, where controlling phonon propagation enhances the efficiency of converting heat into electricity. Phononic crystals, engineered materials with specific phonon band gaps, enable the manipulation of sound and heat waves, paving the way for acoustic and thermal insulation technologies (Draget & Taylor, 2011).

5.3.1.5. Challenges and Future Directions

Despite significant progress, challenges remain in accurately predicting and manipulating phonon properties in complex materials such as alloys and nanostructures. Understanding anharmonic effects, phonon-phonon interactions, and thermal transport across interfaces are active areas of research aimed at improving materials design for energy-efficient devices and advanced thermal management systems.

In conclusion, the normal modes of vibration, represented by phonons in crystalline solids, embody the intricate dynamics of atomic interactions and collective motions within materials. By elucidating these vibrational properties through theoretical models and experimental techniques, researchers continue to unravel the fundamental principles governing thermal and mechanical behaviors in solids, driving innovations across materials science, nanotechnology, and renewable energy applications (Horvath, 2006).

5.3.2. Thermal Properties of Phonons

The thermal properties of phonons play a pivotal role in understanding heat transport mechanisms, thermal conductivity, and thermodynamic behavior in crystalline solids. This section explores the theoretical foundations, mathematical descriptions, and experimental implications of phonon thermal properties, addressing their contributions to material science and engineering (Luck & Petritis, 1986).

5.3.2.1. Phonon Heat Capacity

The heat capacity (C_v) of a material at constant volume is influenced by the contributions from both phonons and electrons. Phonon heat capacity arises from the vibrational modes of atoms in the crystal lattice and can be approximated using the Debye model:

$$C_v = 9Nk_B\left(\frac{T}{\Theta_D}\right)^3 \int_0^{\Theta_D/T} \frac{x^4 e^x}{(e^x - 1)^2} dx$$

where; (N) is the number of atoms in the crystal; (k_B) is the Boltzmann constant; (T) is the temperature; (Θ_D) is the Debye temperature, a characteristic temperature related to the phonon frequencies in the crystal; and $(x = \frac{\hbar\omega}{k_B T})$ with (ω) being the phonon frequency.

This expression provides an estimation of the phonon contribution to the total heat capacity at low temperatures, where phonons dominate thermal transport (O'Leary et al., 1997).

5.3.2.2. Phonon Mean Free Path and Thermal Conductivity

The mean free path (λ) of phonons characterizes the average distance a phonon travels between collisions, influencing thermal conductivity (κ) through the relation:

$$[\kappa = \frac{1}{3} C_v v_{\text{ph}} \lambda]$$

where; (v_{ph}) is the average phonon velocity; and (C_v) is the phonon heat capacity.

Phonon mean free paths are determined by scattering mechanisms such as phonon-phonon interactions, phonon-defect interactions, and boundary scattering.

At high temperatures, Umklapp scattering (inelastic scattering involving momentum transfer) becomes dominant, limiting the mean free path and thereby reducing thermal conductivity (Feng et al., 2015).

5.3.2.3. Phonon Dispersion and Thermal Transport

Phonon dispersion relations, described by the energy as a function of wave vector $(\omega(q))$, govern the propagation of phonons through the crystal lattice. The group velocity $(v_g(q))$ of phonons is related to the slope of the dispersion curve:

$$v_g(q) = \frac{d\omega(q)}{dq}$$

The phonon group velocity influences thermal transport by determining how efficiently phonons carry thermal energy through the material. Materials with higher phonon group velocities generally exhibit higher thermal conductivity (Bocci, 1991).

5.3.2.4. Anharmonic Effects and Thermal Properties

Anharmonic effects, such as phonon-phonon interactions and thermal expansion, significantly impact the thermal properties of materials. The temperature-dependent thermal conductivity $(\kappa(T))$ can be modeled using the Boltzmann transport equation, considering scattering rates and phonon scattering mechanisms across the Brillouin zone. The cumulative effect of anharmonic interactions modifies phonon dispersion relations and affects the temperature dependence of thermal conductivity (Meyer & Richards, 1991).

5.3.2.5. Experimental Techniques and Applications

Experimental techniques like time-domain thermoreflectance, laser flash analysis, and inelastic neutron scattering are used to measure phonon properties and thermal conductivity in solids. These methods provide insights into phonon dispersion relations, mean free paths, and thermal transport mechanisms across a wide range of materials, from semiconductors and metals to complex oxides and nanostructured materials (Thomas et al., 1972).

The understanding of phonon thermal properties is crucial for designing materials with tailored thermal conductivities for applications in thermoelectrics, thermal management, and energy conversion devices. Advances in nanotechnology and materials science enable the engineering of phononic crystals and nanostructures with specific phonon band gaps and controlled thermal properties, offering new avenues for improving energy efficiency and heat dissipation in electronic devices (Nzenwa & Adebayo, 2019).

5.3.2.6. Challenges and Future Directions

Challenges in accurately predicting and manipulating phonon thermal properties include accounting for complex interactions in heterogeneous materials, understanding phonon transport across interfaces, and developing scalable methods for enhancing thermoelectric performance. Future research directions focus on exploring new materials with optimized phonon properties, advancing computational modeling techniques, and innovating experimental methodologies to push the boundaries of thermal materials science.

In conclusion, the thermal properties of phonons underpin the fundamental mechanisms of heat transport and thermodynamic behavior in crystalline solids. By elucidating these properties through theoretical models, experimental measurements, and materials engineering strategies, researchers continue to drive innovations in energy-efficient technologies, thermal management solutions, and sustainable materials for a wide range of industrial applications (Haddon, 1992).

5.3.3. Other Phonon Effects

Phonons, the quantized vibrations of atoms in a crystal lattice, exhibit various effects beyond their thermal properties that profoundly impact the physical behavior and properties of solids. This section explores additional phonon effects, including lattice thermal expansion, phonon-electron interactions, and phonon drag, emphasizing their theoretical foundations, mathematical descriptions, and experimental implications (Heavens et al., 2017).

5.3.3.1. Lattice Thermal Expansion

Lattice thermal expansion refers to the change in dimensions of a crystal lattice in response to temperature variations, primarily driven by phonon vibrations. The linear expansion coefficient (α) relates the change in length (ΔL) of a material to the original length (L_0) and temperature change (ΔT):

$$\frac{\Delta L}{L_0} = \alpha \Delta T$$

The phonon contributions to thermal expansion are described by the Grüneisen parameter (γ), which quantifies the anharmonic effects on phonon frequencies and thermal properties:

$$\gamma = -\frac{V}{\omega}\left(\frac{\partial \omega}{\partial V}\right)_T$$

where; (V) is the volume; and (ω) is the phonon frequency. The Grüneisen parameter characterizes the sensitivity of phonon frequencies to volume changes, influencing the thermal expansion behavior of materials (Fröhlich et al., 1985).

5.3.3.2. Phonon-Electron Interactions

Phonon-electron interactions play a significant role in determining the electrical and thermal transport properties of materials. These interactions can lead to phenomena such as resistivity changes with temperature and phonon drag effects. The scattering rate (τ_{ep}^{-1}) due to phonon-electron interactions affects the electrical resistivity (ρ) through Matthiessen's rule:

$$[\rho(T) = \rho_0 + \rho_{ph}(T) + \rho_{el}(T)]$$

where; is the residual resistivity; $(\rho_{ph}(T))$ is the resistivity due to phonon scattering; and $(\rho_{el}(T))$ the resistivity due to electron-electron interactions. Phonon-electron interactions also contribute to the thermal resistivity of materials, influencing the temperature dependence of thermal conductivity (Rusin & Zawadzki, 2007).

5.3.3.3. Phonon Drag Effect

The phonon drag effect refers to the phenomenon where an electric current induces a drag force on phonons, leading to a net transfer of momentum and energy between electrons and phonons. This effect arises due to the scattering of phonons by moving electrons, resulting in a phonon flow opposite to the electron drift direction.

The phonon drag coefficient (S) quantifies this effect:

$$[j_q = -S j_e]$$

where; (j_q) is the heat flux due to phonon drag; and (j_e) is the electric current density. The phonon drag effect contributes to the thermal conductivity enhancement in metals and semiconductors at low temperatures, impacting the efficiency of thermoelectric devices and thermal management applications (Goldberg et al., 1993).

5.3.3.4. Phonon-Phonon Interactions and Thermal Resistivity

Anharmonic phonon-phonon interactions significantly influence thermal transport properties in materials. At high temperatures, Umklapp scattering processes dominate, limiting the mean free path of phonons and reducing thermal conductivity. The Boltzmann transport equation describes the cumulative effect of phonon scattering mechanisms on thermal resistivity (κ^{-1}):

$$[\kappa^{-1} = \kappa_{ph}^{-1} + \kappa_{el}^{-1}]$$

where; (κ_{ph}^{-1}) and (κ_{el}^{-1})represent the contributions from phonon and electron scattering, respectively. Understanding and controlling phonon-phonon interactions are crucial for designing materials with optimized thermal conductivity for various applications (Zhang & Zhao, 2014).

5.3.3.5. Experimental Techniques and Observations

Experimental techniques such as neutron scattering, Brillouin scattering, and thermal conductivity measurements are employed to investigate phonon effects in solids. Neutron scattering provides detailed information about phonon dispersion relations and phonon lifetimes, offering insights into anharmonic interactions and thermal transport mechanisms. Brillouin scattering enables the characterization of phonon velocities and elastic properties in materials, while thermal conductivity measurements quantify the impact of phonon scattering mechanisms on thermal resistivity (Cao et al., 1997).

Observations of phonon effects in diverse materials, from simple crystals to complex alloys and nanostructures, highlight their role in determining material properties and performance. Advances in experimental methodologies and theoretical modeling continue to advance our understanding of phonon dynamics and facilitate the development of novel materials with tailored thermal, electrical, and mechanical properties (Grimme & Hansen, 2015).

5.3.3.6. Applications and Future Directions

The study of phonon effects underpins advancements in diverse fields, including thermoelectric materials, thermal management technologies, and quantum materials. Tailoring phonon properties through materials design and engineering enables the development of high-performance thermoelectric devices for energy conversion, efficient heat exchangers for thermal management, and phononic crystals for acoustic and thermal insulation (Ingram & Taylor, 1967).

Future research directions focus on exploring novel phonon interactions, manipulating phonon transport properties, and integrating phononic engineering principles into multifunctional materials and devices. Advances in computational modeling, experimental techniques, and materials synthesis are poised to unlock new opportunities for enhancing energy efficiency, sustainability, and technological innovation in the global materials landscape (Somashekar & Joseph, 1996).

In conclusion, the exploration of other phonon effects beyond thermal properties reveals their profound impact on the behavior and functionalities of solids. By elucidating these effects through theoretical frameworks, experimental investigations, and applied research efforts, scientists and engineers continue to advance the frontiers of materials science and pave the way for transformative technologies in the 21st century (Schittekat et al., 2006).

ACTIVITY 5.1.

Explore the normal modes of vibration in crystals using computational simulations or physical models. Investigate how these vibrational modes contribute to the thermal conductivity and other mechanical properties of solids.

SUMMARY

- This chapter delves into the fundamental aspects of solid-state physics, focusing on the microscopic structure of solids and their various properties. It begins with an exploration of lattices and crystal structures, where the arrangement of atoms in a solid is described in a periodic and symmetrical manner. The concept of the reciprocal lattice is introduced, providing a crucial framework for understanding diffraction phenomena and the behavior of waves in periodic structures.
- The chapter then covers diffraction experiments, which are pivotal in determining the atomic structure of crystals. Techniques such as X-ray diffraction and electron diffraction are discussed, highlighting their importance in revealing the detailed arrangement of atoms within a solid.
- Following this, the behavior of electrons in crystals is examined through models of electronic bands. The formation and filling of these bands are crucial for understanding the electrical properties of materials, distinguishing between metals, insulators, and semiconductors. The spectra of electrons in solids, including the density of states and electronic transitions, are also analyzed.
- The vibrations of crystals are another critical area of study. The chapter discusses the normal modes of vibration, known as phonons, which play a significant role in the thermal and acoustic properties of solids. The thermal properties of phonons, including heat capacity and thermal conductivity, are examined, along with other phonon effects such as thermal expansion and interaction with electrons.

REVIEW QUESTIONS

1. Explain the concept of reciprocal lattice and its significance in solid-state physics.
2. How do diffraction experiments contribute to our understanding of crystal structures? Provide examples.
3. Describe the main types of electronic bands in crystals and their distinguishing characteristics.
4. What are phonons, and how do they contribute to the thermal properties of solids?
5. Discuss the differences between metals, insulators, and semiconductors based on their electronic band structures.

MULTIPLE CHOICE QUESTIONS

1. Which experimental technique is commonly used to determine the atomic structure of crystals?

a. NMR spectroscopy

b. X-ray diffraction

c. Mass spectrometry

d. Infrared spectroscopy

2. The concept of reciprocal lattice is crucial for understanding:

a. Phonon dynamics

b. Crystal growth

c. Diffraction patterns

d. Magnetic properties

3. Which of the following materials typically has a completely filled valence band and an empty conduction band?

a. Metal

b. Semiconductor

c. Insulator

d. Superconductor

4. Phonons in solids primarily refer to:

a. Electromagnetic waves

b. Vibrations of crystal lattice

c. Free electrons

d. Optical transitions

5. Thermal conductivity in solids is mainly influenced by:

a. Electronic band structure

b. Phonon scattering

c. Magnetic interactions

d. Optical properties

Answers to Multiple Questions

1. (b); 2. (c); 3. (c); 4. (b); 5. (b).

REFERENCES

1. Apweiler, R., Aslanidis, C., Deufel, T., Gerstner, A., Hansen, J., Hochstrasser, D., & Schmitz, G. (2009). Approaching clinical proteomics: Current state and future fields of application in fluid proteomics. *Clinical Chemistry and Laboratory Medicine, 47*(6), 724–744.
2. Balandin, A. A. (2005). Nanophononics: Phonon engineering in nanostructures and nanodevices. *Journal of Nanoscience and Nanotechnology, 5*(7), 1015–1022.
3. Båth, M. (2010). Evaluating imaging systems: Practical applications. *Radiation Protection Dosimetry, 139*(1–3), 26–36.

4. Bocci, V. (1991). Interleukins: Clinical pharmacokinetics and practical implications. *Clinical Pharmacokinetics, 21*(4), 274–284.
5. Bresler, E., & Dagan, G. (1983). Unsaturated flow in spatially variable fields: 2. Application of water flow models to various fields. *Water Resources Research, 19*(2), 421–428.
6. Byrne, C., Subramanian, G., & Pillai, S. C. (2018). Recent advances in photocatalysis for environmental applications. *Journal of Environmental Chemical Engineering, 6*(3), 3531–3555.
7. Cahill, D. G., & Pohl, R. O. (1988). Lattice vibrations and heat transport in crystals and glasses. *Annual Review of Physical Chemistry, 39*(1), 93–121.
8. Cao, T. H., Creasy, P. N., & Wuwongse, V. (1997). Fuzzy types and their lattices. In *Proceedings of the 6th International Fuzzy Systems Conference* (Vol. 2, pp. 805–812).
9. Daniels, J., Festenberg, C. V., Raether, H., & Zeppenfeld, K. (1970). Optical constants of solids by electron spectroscopy. *Springer Tracts in Modern Physics, 54*, 77–135.
10. Datta, S., & Grant, D. J. (2004). Crystal structures of drugs: Advances in determination, prediction, and engineering. *Nature Reviews Drug Discovery, 3*(1), 42–57.
11. Davisson, C. J. (1928). The diffraction of electrons by a crystal of nickel. *Bell System Technical Journal, 7*(1), 90–105.
12. Diamond, R. (1971). A real-space refinement procedure for proteins. *Acta Crystallographica Section A: Crystal Physics, Diffraction, Theoretical and General Crystallography, 27*(5), 436–452.
13. Dorrian, J. F., Newnham, R. E., Smith, D. K., & Kay, M. I. (1972). Crystal structure of Bi⊠Ti⊠O⊠⊠. *Ferroelectrics, 3*(1), 17–27.
14. Draget, K. I., & Taylor, C. (2011). Chemical, physical and biological properties of alginates and their biomedical implications. *Food Hydrocolloids, 25*(2), 251–256.
15. Drummond, N. D., Zolyomi, V., & Fal'Ko, V. I. (2012). Electrically tunable band gap in silicene. *Physical Review B—Condensed Matter and Materials Physics, 85*(7), 075423.
16. Ebalard, S., & Spaepen, F. (1989). The body-centered-cubic-type icosahedral reciprocal lattice of the Al–Cu–Fe quasi-periodic crystal. *Journal of Materials Research, 4*(1), 39–43.
17. Epstein, G. (1960). The lattice theory of Post algebras. *Transactions of the American Mathematical Society, 95*(2), 300–317.
18. Feng, T., Ruan, X., Ye, Z., & Cao, B. (2015). Spectral phonon means free path and thermal conductivity accumulation in defected graphene: The effects of defect type and concentration. *Physical Review B, 91*(22), 224301.
19. Field, D. P. (1997). Recent advances in the application of orientation imaging. *Ultramicroscopy, 67*(1–4), 1–9.
20. Field, J. E., Walley, S. M., Proud, W. G., Goldrein, H. T., & Siviour, C. R. (2004). Review of experimental techniques for high-rate deformation and shock studies. *International Journal of Impact Engineering, 30*(7), 725–775.

21. Fröhlich, J., Martinelli, F., Scoppola, E., & Spencer, T. (1985). Constructive proof of localization in the Anderson tight binding model. *Communications in Mathematical Physics, 101*(1), 21–46.
22. Goldberg, R. B., Beals, T. P., & Sanders, P. M. (1993). Anther development: Basic principles and practical applications. *The Plant Cell, 5*(10), 1217.
23. Grimme, S., & Hansen, A. (2015). A practicable real-space measure and visualization of static electron-correlation effects. *Angewandte Chemie International Edition, 54*(42), 12308–12313.
24. Gruber, B. (1997). Classification of lattices: A new step. *Acta Crystallographica Section A: Foundations of Crystallography, 53*(4), 505–521.
25. Gu, B., Zhao, C., Baev, A., Yong, K. T., Wen, S., & Prasad, P. N. (2016). Molecular nonlinear optics: Recent advances and applications. *Advances in Optics and Photonics, 8*(2), 328–369.
26. Haddon, R. C. (1992). Electronic structure, conductivity and superconductivity of alkali metal doped (C_{60}). *Accounts of Chemical Research, 25*(3), 127–133.
27. Heavens, A., Fantaye, Y., Sellentin, E., Eggers, H., Hosenie, Z., Kroon, S., & Mootoovaloo, A. (2017). No evidence for extensions to the standard cosmological model. *Physical Review Letters, 119*(10), 101301.
28. Horvath, A. O. (2006). The alliance in context: Accomplishments, challenges, and future directions. *Psychotherapy: Theory, Research, Practice, Training, 43*(3), 258.
29. Hotelling, H. (1944). Some improvements in weighing and other experimental techniques. *The Annals of Mathematical Statistics, 15*(3), 297–306.
30. Ingram, L., & Taylor, H. F. W. (1967). The crystal structures of sjögrenite and pyroaurite. *Mineralogical Magazine and Journal of the Mineralogical Society, 36*(280), 465–479.
31. Issendorff, B. V., & Cheshnovsky, O. (2005). Metal to insulator transitions in clusters. *Annual Review of Physical Chemistry, 56*(1), 549–580.
32. Kehr, J. (1993). A survey on quantitative micro dialysis: Theoretical models and practical implications. *Journal of Neuroscience Methods, 48*(3), 251–261.
33. Khazaei, M., Ranjbar, A., Ghorbani-Asl, M., Arai, M., Sasaki, T., Liang, Y., & Yunoki, S. (2016). Nearly free electron states in MXenes. *Physical Review B, 93*(20), 205125.
34. Kosobutskii, A. V., & Poplavnoi, A. S. (2006). Modeling band spectra of compound crystals in the basis of states of their sublattices. *Russian Physics Journal, 49*(7), 772–777.
35. Kurbatsky, V. P., Korotun, A. V., Babich, A. V., & Pogosov, V. V. (2009). Fermi energy and optical conductivity of metal quantum wires. *Physics of the Solid State, 51*, 2520–2528.
36. Li, F., Whalley, J., & Williams, H. (2001). Between physical and electronic spaces: The implications for organisations in the networked economy. *Environment and Planning A, 33*(4), 699–716.

37. Littlewood, P. B., Mihaila, B., Schulze, R. K., Safarik, D. J., Gubernatis, J. E., Bostwick, A., & Lashley, J. C. (2010). Band structure of SnTe studied by photoemission spectroscopy. *Physical Review Letters, 105*(8), 086404.

38. Lucas-González, R., Viuda-Martos, M., Pérez-Alvarez, J. A., & Fernández-López, J. (2018). *In vitro* digestion models suitable for foods: Opportunities for new fields of application and challenges. *Food Research International, 107*, 423–436.

39. Luck, J. M., & Petritis, D. (1986). Phonon spectra in one-dimensional quasicrystals. *Journal of Statistical Physics, 42*, 289–310.

40. Meyer, A. Y., & Richards, W. G. (1991). Similarity of molecular shape. *Journal of Computer-Aided Molecular Design, 5*, 427–439.

41. Mondal, S. (2017). Preparation, properties and applications of nanocellulosic materials. *Carbohydrate Polymers, 163*, 301–316.

42. Nzenwa, E. C., & Adebayo, A. D. (2019). Analysis of insulators for distribution and transmission networks. *American Journal of Engineering Research (AJER), 8*(12), 138–145.

43. O'Leary, S. K., Johnson, S. R., & Lim, P. K. (1997). The relationship between the distribution of electronic states and the optical absorption spectrum of an amorphous semiconductor: An empirical analysis. *Journal of Applied Physics, 82*(7), 3334–3340.

44. Pasinetti, L. L. (1960). A mathematical formulation of the Ricardian system. *The Review of Economic Studies, 27*(2), 78–98.

45. Paukstelis, P. J., Nowakowski, J., Birktoft, J. J., & Seeman, N. C. (2004). Crystal structure of a continuous three-dimensional DNA lattice. *Chemistry & Biology, 11*(8), 1119–1126.

46. Pendás, A. M., Blanco, M. A., & Francisco, E. (2007). Chemical fragments in real space: Definitions, properties, and energetic decompositions. *Journal of Computational Chemistry, 28*(1), 161–184.

47. Pina, J. M., Martinez, E., De Chernatony, L., & Drury, S. (2006). The effect of service brand extensions on corporate image: An empirical model. *European Journal of Marketing, 40*(1/2), 174–197.

48. Ridolfi, E., Le, D., Rahman, T. S., Mucciolo, E. R., & Lewenkopf, C. H. (2015). A tight-binding model for MoS⊠ monolayers. *Journal of Physics: Condensed Matter, 27*(36), 365501.

49. Rota, G. C. (1997). The many lives of lattice theory. *Notices of the AMS, 44*(11), 1440–1445.

50. Rusin, T. M., & Zawadzki, W. (2007). Zitterbewegung of nearly-free and tightly-bound electrons in semiconductors. *Journal of Physics: Condensed Matter, 19*(13), 136219.

51. Schittekat, P., Sevaux, M., & Sorensen, K. (2006). A mathematical formulation for a school bus routing problem. In *2006 International Conference on Service Systems and Service Management* (Vol. 2, pp. 1552–1557).

52. Shaw, S. W., & Pierre, C. (1993). Normal modes for non-linear vibratory systems. *Journal of Sound and Vibration, 164*(1), 85–124.

53. Shvyd'Ko, Y., & Lindberg, R. (2012). Spatiotemporal response of crystals in x-ray Bragg diffraction. *Physical Review Special Topics—Accelerators and Beams, 15*(10), 100702.

54. Sivarajah, P., Maznev, A. A., Ofori-Okai, B. K., & Nelson, K. A. (2016). What is the Brillouin zone of an anisotropic photonic crystal? *Physical Review B, 93*(5), 054204.

55. Smith, K. C. A., & Oatley, C. W. (1955). The scanning electron microscope and its fields of application. *British Journal of Applied Physics, 6*(11), 391.

56. Somashekar, D., & Joseph, R. (1996). Chitosanases—Properties and applications: A review. *Bioresource Technology, 55*(1), 35–45.

57. Stokes, H. T., & Hatch, D. M. (2002). Procedure for obtaining microscopic mechanisms of reconstructive phase transitions in crystalline solids. *Physical Review B, 65*(14), 144114.

58. Stutz, J., & Platt, U. (1996). Numerical analysis and estimation of the statistical error of differential optical absorption spectroscopy measurements with least-squares methods. *Applied Optics, 35*(30), 604.

59. Thomas, J. M., Evans, E. L., & Williams, J. O. (1972). Microscopic studies of enhanced reactivity at structural faults in solids. *Proceedings of the Royal Society of London. A. Mathematical and Physical Sciences, 331*(1586), 417–427.

60. Ungár, T. (2001). Dislocation densities, arrangements and character from X-ray diffraction experiments. *Materials Science and Engineering: A, 309*, 14–22.

61. Weigel, D. (1983). Bragg's law in higher dimensions. *Acta Crystallographic Section A: Foundations of Crystallography, 39*(1), 76–78.

62. Zeiger, H. J., Vidal, J., Cheng, T. K., Ippen, E. P., Dresselhaus, G., & Dresselhaus, M. S. (1992). Theory for displacive excitation of coherent phonons. *Physical Review B, 45*(2), 768.

63. Zhang, Q. B., & Zhao, J. (2014). A review of dynamic experimental techniques and mechanical behavior of rock materials. *Rock Mechanics and Rock Engineering, 47*, 1411–1478.

CHAPTER

6 Liquids and Gases

LEARNING OBJECTIVES

After studying this chapter, you will be able to:

1. Understand the molecular structure and interactions in liquids and gases;
2. Describe the properties of liquids, including density, compressibility, viscosity, and surface tension;
3. Explain the behavior of gases through the ideal gas law and kinetic theory;
4. Analyze deviations from ideal gas behavior and the equations of state for real gases;
5. Understand phase transitions and their thermodynamic implications;
6. Apply the Gibbs phase rule and chemical potential concepts to phase equilibria;
7. Relate thermodynamic equilibria to atmospheric physics and climate applications.

KEY TERMS FROM THIS CHAPTER

Compressibility
Ideal gas law
Molecular interactions
Surface tension
Gibbs phase rule
Kinetic theory
Phase transitions
Viscosity

UNIT INTRODUCTION

Liquids and gases are fundamental states of matter that play pivotal roles in understanding the physical properties and behaviors of substances. In the study of physics of matter, liquids are characterized by their ability to flow and take the shape of their container, exhibiting properties such as viscosity, surface tension, and density. The cohesive forces between molecules in liquids contribute to their unique behavior, influencing phenomena like capillary action and meniscus formation. Understanding these properties is crucial in fields ranging from chemistry to engineering, as they dictate how substances interact at a molecular level and affect processes such as diffusion and solvent behavior in solutions (Loitsyanskiy, 1966).

Similarly, gases, characterized by their ability to expand to fill any container uniformly, are governed by principles outlined in the kinetic theory of gases and the ideal gas law. Their behavior is heavily influenced by factors such as temperature, pressure, and volume, which are interrelated through mathematical expressions that describe their state. Real gases, which deviate from ideal behavior under certain conditions, require more complex equations of state to accurately describe their properties. The study of gases is vital in fields like atmospheric physics, where understanding gas behavior aids in predicting weather patterns and studying climate change.

In both liquids and gases, phase transitions—such as vaporization, condensation, melting, and freezing—are critical phenomena that elucidate how matter transforms under different conditions of temperature and pressure (Jaeger et al., 2018). These transitions are essential in fields as diverse as materials science, where controlling phase changes enables the design of new materials with specific properties, and environmental science, where understanding water's phase transitions informs models of Earth's water cycle.

Overall, the study of liquids and gases within the framework of "Physics of Matter" provides essential insights into the fundamental behaviors of matter at the atomic and molecular levels. Their properties and transformations underpin numerous scientific disciplines and technological applications, making them foundational topics in the pursuit of understanding and manipulating matter for societal and industrial advancements.

6.1. PROPERTIES AND BEHAVIOR OF LIQUIDS

The properties and behavior of liquids encompass a diverse range of physical characteristics that distinguish them from other states of matter. Liquids are defined by their ability to flow and take on the shape of their container, due to weak intermolecular forces that allow molecules to move past each other while maintaining some degree of cohesion. Key properties include viscosity, which measures resistance to flow, and surface tension, which governs interactions at the liquid's interface. Density, a measure of mass per unit volume, influences buoyancy and mixing behaviors (Mao et al., 2018). Understanding these properties is crucial in fields like chemistry, where they dictate reactions and solubility, and in engineering, where they inform design considerations for fluid dynamics and materials processing.

6.1.1. Molecular Structure and Interactions

In the realm of liquids, understanding molecular structure and interactions is fundamental to comprehending their physical properties and behaviors. Liquids are characterized by molecules that maintain a degree of mobility while exhibiting cohesive forces that keep them together.

The molecular structure of a liquid is defined by the spatial arrangement of its constituent molecules and the types of chemical bonds that hold them together. For instance, in water (H_2O), molecules are held together by polar covalent bonds between hydrogen and oxygen atoms. This polar nature gives rise to hydrogen bonding, a strong intermolecular force that significantly influences the properties of liquid water (Figure 6.1).

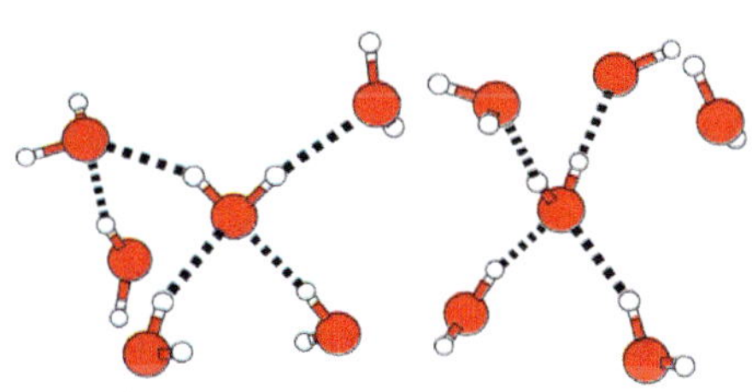

(a) (b)

Figure 6.1. *Snapshots of local structures in liquid water.*

Source: Soohaeng Yoo Willow, Creative Commons License.

Mathematically, these interactions can be described using potential energy functions that quantify the energy associated with different molecular configurations. For example, the Lennard-Jones potential $V(r)$ is often used to model the intermolecular interactions in liquids, where (r) represents the distance between molecules (Hossain & Davies, 2013):

Remember

Remember that the properties and behavior of liquids and gases are governed by fundamental principles of molecular interactions and kinetic theory, influencing their physical and chemical characteristics.

$$V(r) = 4\epsilon\left[\left(\frac{\sigma}{r}\right)^{12} - \left(\frac{\sigma}{r}\right)^{6}\right]$$

where; (ϵ) is the depth of the potential well; and (σ) is the distance at which the potential energy between molecules is zero.

The nature of these interactions dictates several key properties of liquids. Viscosity, for example, is influenced by the strength of intermolecular forces—liquids with stronger cohesive forces exhibit higher viscosity due to increased resistance to flow. Surface tension, another critical property, arises from the imbalance of intermolecular forces at the liquid's surface, leading to a tendency to minimize surface area.

Furthermore, the molecular arrangement in liquids affects their density, with closely packed arrangements resulting in higher densities. This parameter plays a crucial role in determining the behavior of liquids under different temperature and pressure conditions, as described by the equation of state specific to each substance.

Overall, a detailed understanding of molecular structure and interactions in liquids involves rigorous mathematical modeling and empirical observations. This knowledge forms the basis for predicting and manipulating the physical properties of liquids across various scientific and industrial applications, from materials science to pharmaceuticals (Kauzmann, 1948).

6.1.2. Density and Compressibility

Density and compressibility are fundamental properties that characterize the physical behavior of liquids. Understanding these properties requires a grasp of how molecular interactions and external conditions influence the molecular arrangement and volume of a liquid.

6.1.2.1. Density

Density (ρ) is defined as the mass per unit volume of a substance and is a crucial parameter in describing its compactness and mass distribution. Mathematically, density is expressed as:

$$\rho = \frac{m}{V}$$

where; (m) is the mass of the liquid; and (V) is its volume. In liquids, density varies with temperature and pressure due to changes in molecular spacing and intermolecular forces (Domańska, 2006). For instance, as temperature increases, molecules gain kinetic energy, causing them to vibrate more vigorously and thereby increasing the average intermolecular distance, which generally decreases density. However, this trend is not universal, as water exhibits anomalous behavior by contracting as it cools below 4°C before expanding upon freezing.

6.1.2.2. Compressibility

Compressibility (β) measures the extent to which a substance can be compressed under the application of external pressure. In liquids, compressibility is typically very low compared to gases due to the strong intermolecular forces that resist volume reduction. The inverse of compressibility, the bulk modulus (K), relates pressure (P) to fractional volume change $(\Delta V/V)$:

$$K = -V\left(\frac{\partial P}{\partial V}\right)_T$$

where; $\left(\frac{\partial P}{\partial V}\right)_T$

denotes the isothermal pressure-volume derivative.

For liquids, $(K)$is often related to the cohesive energy density and reflects the resistance of the liquid to compression. Liquids with higher densities typically have lower compressibilities and vice versa (Rooney et al., 2010).

6.1.3. Viscosity and Surface Tension

Viscosity and surface tension are crucial properties of liquids that govern their flow behavior and interactions at interfaces. Understanding these phenomena involves delving into molecular interactions and the physical forces that define liquid behavior (Rebelo et al., 2005).

6.1.3.1. Viscosity

Viscosity (η) is a measure of a liquid's resistance to flow. It arises due to internal frictional forces between adjacent layers of molecules as they move past each other. Mathematically, viscosity is defined by the ratio of shear stress (τ) to shear rate $(\dot{\gamma})$:

$$\tau = \eta\dot{\gamma}$$

where; $(\dot{\gamma})$represents the velocity gradient across the liquid.

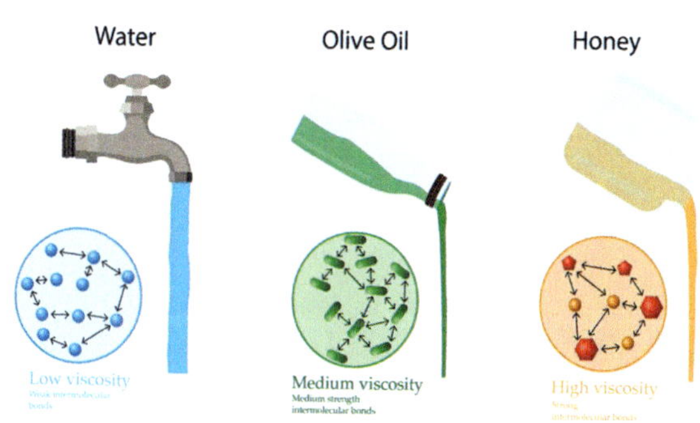

Figure 6.2. *Comparison of viscosity of different liquids.*

Source: Thermo Blanket, Creative Commons License.

For Newtonian fluids, such as water and most common liquids, viscosity remains constant regardless of the applied shear stress. Non-Newtonian fluids, however, exhibit variable viscosity depending on the applied stress conditions (Figure 6.2). The molecular origin of viscosity lies in intermolecular forces, primarily van der Waals forces and hydrogen bonding in polar liquids. Stronger intermolecular interactions result in higher viscosity due to increased resistance to molecular motion (Johansson & Curstedt, 1997).

6.1.3.2. Surface Tension

Surface tension (γ) is the energy required to increase the surface area of a liquid by a unit area. It is caused by the cohesive forces between molecules at the liquid-air interface. Mathematically, surface tension is given by:

$$\gamma = \frac{F}{l}$$

where; (F) is the force acting parallel to the surface; and (l) is the length along which the force acts.

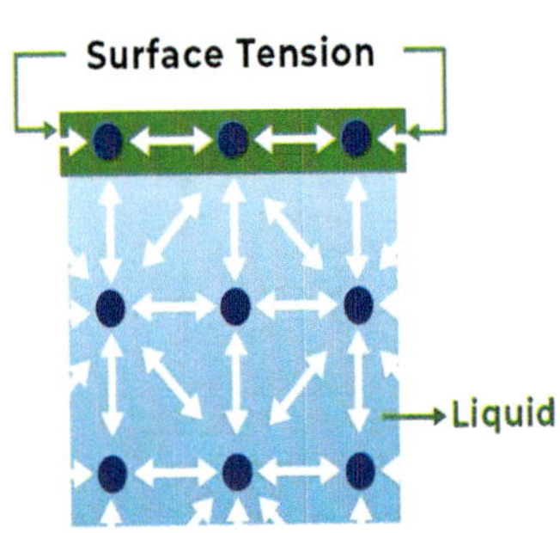

Figure 6.3. *Illustration of surface tension of liquids.*

Source: Geeks for Geeks, Creative Commons License.

Surface tension tends to minimize the surface area of a liquid, leading to phenomena such as the formation of droplets and capillary rise. The magnitude of surface tension depends on the nature of the molecules and their interactions. For example, water exhibits high surface tension due to strong hydrogen bonding at its surface (Figure 6.3).

6.1.4. Capillary Action and Meniscus Formation

Capillary action and meniscus formation are phenomena that arise from the interaction between liquids and solid surfaces, driven by adhesive and cohesive forces at play (Fleischer & Payne, 2014).

6.1.4.1. Capillary Action

Capillary action refers to the ability of a liquid to flow in narrow spaces or tubes against the force of gravity. It occurs due to the combination of adhesive forces (between liquid and solid) and cohesive forces (within the liquid itself). The height (h) to which a liquid rises or falls in a capillary tube can be described by the Jurin's law equation:

$$h = \frac{2\gamma \cos\theta}{\rho g r}$$

where; (γ) is the surface tension of the liquid; (θ) is the contact angle between the liquid and the surface of the tube; (ρ) is the density of the liquid; (g) is the acceleration due to gravity; and (r) is the radius of the capillary tube.

Capillary action is significant in biological systems, such as in plants where it facilitates the movement of water from roots to leaves, and in medical diagnostics where it aids in fluid movement in microfluidic devices.

6.1.4.2. Meniscus Formation

The meniscus is the curve seen at the surface of a liquid in response to its container. It forms due to the adhesive forces between the liquid and the container walls, which differ depending on whether the liquid wets or does not wet the container surface (Chothia & Jones, 1997).

Mathematically, the shape of the meniscus is described by the Young-Laplace equation:

$$P = P_{\text{inside}} - P_{\text{outside}} = \frac{2\gamma}{R}$$

where; (P) is the pressure difference across the meniscus; (γ) is the surface tension of the liquid; (R) is the radius of curvature of the meniscus.

The curvature of the meniscus affects phenomena such as capillary rise and the accuracy of liquid volume measurements in instruments like pipettes and burettes.

In conclusion, capillary action and meniscus formation illustrate the intricate balance between adhesive and cohesive forces in liquids (Kim et al., 2000). Their mathematical foundations allow for precise predictions and applications across various scientific disciplines, from materials science to engineering and medicine.

6.2. PROPERTIES AND BEHAVIOR OF GASES

The properties and behavior of gases are characterized by their ability to expand to fill any container uniformly and to be compressed under pressure. Gases do not have a fixed shape or volume, instead conforming to the shape and volume of their container due to their particles having significant kinetic energy and negligible intermolecular forces. These particles move randomly and collide elastically with each other and the container walls, leading to properties like high compressibility and low density compared to liquids and solids (Fredlund, 1976). Understanding these behaviors is essential in fields such as thermodynamics, atmospheric science, and engineering, where gas laws and behaviors dictate processes ranging from combustion engines to weather patterns.

6.2.1. Ideal Gas Law and Deviations

The ideal gas law is a fundamental equation in the study of gases, describing the relationship between pressure (P), volume (V), and temperature (T) for an ideal gas. It is expressed as:

$$PV = nRT$$

where; (P) is the pressure of the gas; (V) is the volume it occupies; (n) is the number of moles of gas; (R) is the ideal gas constant $(R = 8.314\,\text{J/mol·K})$; and (T) is the absolute temperature.

The ideal gas law assumes several idealized conditions:

1. **Particles have No Volume:** The volume occupied by the gas molecules themselves is negligible compared to the volume of the container.
2. **No Intermolecular Forces:** Molecules do not attract or repel each other; they only interact through elastic collisions.
3. **Perfect Elasticity:** Collisions between gas molecules and container walls are perfectly elastic, conserving both momentum and kinetic energy.

Deviation from ideal gas behavior occurs under conditions where these assumptions no longer hold true, such as at high pressures or low temperatures. These deviations are quantitatively described by the compressibility factor (Z), defined as the ratio of the actual volume occupied by the gas to the volume predicted by the ideal gas law (Elsharkawy, 2004):

$$Z = \frac{PV}{nRT}$$

For real gases, (Z) deviates from 1 under conditions where intermolecular forces become significant or when molecular volumes are non-negligible. Corrections to the ideal gas law, such as the van der Waals equation, account for these deviations:

$$\left(P + \frac{a}{V_m^2}\right)(V_m - b) = RT$$

where; (a) adjusts for attractive forces between molecules; and (b) adjusts for molar volume, usually the volume per mole of gas.

While the ideal gas law provides a useful approximation under many conditions, understanding deviations from ideal behavior is crucial for accurately modeling and predicting gas properties in diverse practical applications.

6.2.2. Kinetic Theory of Gases

The kinetic theory of gases provides a theoretical framework to explain the macroscopic properties of gases based on the behavior of their constituent molecules. Key principles of the kinetic theory include:

1. **Molecular Motion:** Gas molecules move randomly and in straight lines between collisions with each other and the container walls.
2. **Elastic Collisions:** Collisions between gas molecules and with the container walls are perfectly elastic, conserving both momentum and kinetic energy.
3. **Negligible Volume:** The volume occupied by gas molecules themselves is negligible compared to the volume of the container.
4. **Temperature:** The temperature of a gas is directly proportional to the average kinetic energy of its molecules.

The kinetic theory allows derivation of macroscopic gas laws from microscopic principles. For instance, the relationship between pressure and kinetic theory can be derived from the momentum change per unit area during collisions:

Tip

According to kinetic theory, the pressure exerted by a gas is related to the average kinetic energy of its molecules and the frequency of their collisions with the container walls. This theory provides a molecular-level explanation for gas behavior.

$$P = \frac{1}{3} nm\overline{v^2}$$

where; (n) is the number of molecules per unit volume; (m) is the mass of a gas molecule; and $\left(\overline{v^2}\right)$ is the mean square speed of the molecules.

The theory also explains the origin of temperature and the distribution of molecular speeds. The Maxwell-Boltzmann distribution describes the distribution of molecular speeds in a gas at a given temperature (Norton, 1983):

$$f(v) = 4\pi \left(\frac{m}{2\pi k_B T} \right)^{3/2} v^2 e^{-\frac{mv^2}{2k_B T}}$$

where; $f(v)$ is the fraction of molecules with speed (v); and (k_B) is Boltzmann's constant $\left(k_B = 1.381 \times 10^{-23}\ \text{J/K}\right)$.

The kinetic theory provides a robust framework for understanding gas behavior across a wide range of conditions, from ideal gases to real gases and under various temperatures and pressures.

6.2.3. Real Gases and Equation of State

Real gases deviate from ideal gas behavior due to intermolecular forces and finite molecular volumes. Several equations of state have been developed to describe real gases more accurately than the ideal gas law alone:

1. **van der Waals Equation:** The van der Waals equation incorporates corrections for both molecular volume and intermolecular forces (Gaiduk et al., 2015):

$$\left(P + \frac{an^2}{V^2} \right)(V - nb) = nRT$$

where; (a) adjusts for attractive forces between molecules; (b) adjusts for molecular volume; (n) is the number of moles of gas; and (V) is the volume occupied by the gas.

2. **Redlich-Kwong Equation:** The Redlich-Kwong equation provides another approach to correct for molecular volume while assuming a weaker dependence on intermolecular forces:

$$P = \frac{RT}{V_m - b} - \frac{a}{\sqrt{T} V_m (V_m + b)}$$

where; (V_m) is the molar volume of the gas; and (a) and (b) are constants specific to the gas.

These equations improve accuracy over a wider range of temperatures and pressures than the ideal gas law alone. They are derived from statistical mechanics principles and experimental data, adjusting for real gas behavior observed under various conditions.

6.2.4. Gas Mixtures and Partial Pressures

Gas mixtures consist of two or more gases occupying the same space. The behavior of each gas in a mixture is independent of the others, governed by Dalton's Law of Partial Pressures (Ranga & Asawa, 1977):

$$P_{\text{total}} = P_1 + P_2 + \ldots + P_n$$

where; (P_{total}) is the total pressure of the mixture; and (P_i) is the partial pressure of gas (i).

Each gas in a mixture behaves as if it occupies the entire volume alone, exerting a pressure proportional to its mole fraction and temperature. The partial pressure of a gas can be calculated using its mole fraction (x_i) and the total pressure:

$$P_i = x_i P_{\text{total}}$$

where; $\left(x_i = \frac{n_i}{n_{\text{total}}}\right)$

is the mole fraction of gas (i) in the mixture.

Understanding partial pressures is crucial in various applications, such as in gas-phase reactions, where the rate of reaction depends on the partial pressures of reactants, and in gas mixtures used in industrial processes and environmental monitoring (Khattab et al., 2012). In conclusion, these topics provide a comprehensive understanding of gas behavior, from ideal to real gases, including their mixture properties and deviations from ideal behavior under different conditions. The mathematical foundations and theoretical frameworks outlined enable precise modeling and prediction of gas properties crucial in scientific research, engineering design, and technological applications.

6.3. PHASE TRANSITIONS

Phase transitions are fundamental changes in the physical state of matter, such as from solid to liquid (melting), liquid to gas (vaporization), or the reverse processes (freezing and condensation). These transitions occur due to alterations in temperature or pressure, causing shifts in the arrangement and energy state of molecules or atoms within a substance. At specific conditions known as phase equilibrium, two phases coexist in equilibrium. Understanding phase transitions is crucial in fields ranging from material sciences to atmospheric physics, influencing the design of materials, energy transfer processes, and environmental dynamics (Figure 6.4).

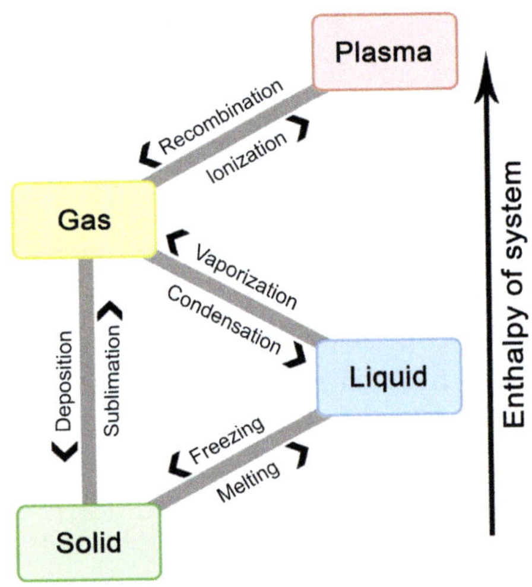

Figure 6.4. *This diagram shows the nomenclature for the different phase transitions.*

Source: Flanker, Creative Commons License.

6.3.1. Vaporization and Condensation

Vaporization, or evaporation, is the phase transition from the liquid phase to the gas phase. It occurs when molecules at the liquid surface gain enough kinetic energy to overcome the intermolecular forces holding them in the liquid phase. The rate of vaporization increases with temperature and surface area and is influenced by factors such as humidity and atmospheric pressure. Mathematically, the rate of vaporization can be described by the Clausius-Clapeyron equation:

$$\frac{dP}{dT} = \frac{\Delta H_{\text{vap}}}{T\Delta V}$$

where; (ΔH_{vap}) is the enthalpy of vaporization; (T) is the temperature; and (ΔV) is the change in volume during vaporization.

Condensation is the reverse process of vaporization, where gas molecules lose energy (usually through cooling) and transition into the liquid phase. It occurs when the vapor pressure of the gas equals the equilibrium vapor pressure of the liquid at a given temperature. Condensation plays a crucial role in weather phenomena such as cloud formation and is essential in processes like distillation and refrigeration (Pelofsky, 1966).

6.3.2. Melting and Freezing

Melting, or fusion, is the phase transition from the solid phase to the liquid phase.

It occurs when the thermal energy supplied to a solid substance increases the kinetic energy of its particles to overcome the forces holding them in a fixed lattice structure. The melting point of a substance is the temperature at which solid and liquid phases coexist in equilibrium. For pure substances, the melting point remains constant under atmospheric pressure.

Freezing is the reverse process of melting, where a liquid substance loses thermal energy and transitions into the solid phase.

The freezing point of a substance is identical to its melting point, representing the temperature at which liquid and solid phases coexist in equilibrium. The process of freezing is critical in food preservation, cryopreservation, and the formation of ice in natural and engineered systems.

6.3.3. Sublimation and Deposition

Sublimation is the phase transition directly from the solid phase to the gas phase without passing through the intermediate liquid phase. It occurs when the vapor pressure of the solid exceeds atmospheric pressure at a given temperature. Examples include dry ice (solid carbon dioxide) and mothballs (naphthalene). The rate of sublimation depends on temperature and pressure conditions (Figure 6.5).

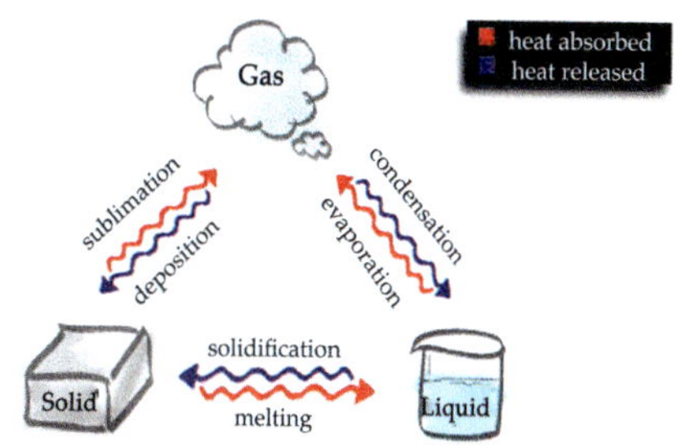

Figure 6.5. *Illustration of changes in phases (sublimation and deposition).*

Source: Stack Exchange, Creative Commons License.

Deposition is the reverse process of sublimation, where gas molecules lose energy and transition directly into the solid phase. It occurs when the equilibrium vapor pressure of the substance is lower than the atmospheric pressure at a given temperature (Brooks et al., 2001). Deposition is involved in phenomena such as frost formation on surfaces and the formation of snowflakes in clouds.

6.3.4. Critical Points and Phase Diagrams

The critical point of a substance is the temperature and pressure at which the liquid and gas phases of the substance become indistinguishable. At the critical point, the liquid and gas phases have the same density, and the distinction between phases disappears. Above the critical temperature $((T_c))$ and critical pressure $((P_c))$, the substance exists as a supercritical fluid, exhibiting properties of both liquids and gases. The critical point is crucial in understanding phase behavior under extreme conditions and in industrial processes like supercritical fluid extraction.

Phase diagrams are graphical representations that depict the equilibrium conditions between different phases of a substance as a function of temperature and pressure. They illustrate the boundaries between solid, liquid, and gas phases and provide insights into phase transitions and critical points. Key features of a phase diagram include (Sahasrabudhe et al., 2017):

i. **Triple Point:** The temperature and pressure at which all three phases (solid, liquid, and gas) coexist in equilibrium.

ii. **Phase Boundaries:** Lines that separate different phases (solid-liquid, liquid-gas, solid-gas).

iii. **Critical Point:** The endpoint of the liquid-gas phase boundary.

Phase diagrams are essential tools in materials science, thermodynamics, and engineering, guiding the development of new materials, understanding the behavior of mixtures, and optimizing industrial processes such as alloy manufacturing and phase separation in chemical reactions.

In conclusion, understanding vaporization, condensation, melting, freezing, sublimation, deposition, critical points, and phase diagrams provides a comprehensive framework for comprehending the behavior of substances under different conditions. These concepts are fundamental in various scientific disciplines and practical applications, from designing materials to controlling environmental processes and optimizing industrial operations.

Did you know?

Did you know that critical points in phase diagrams represent the conditions where a substance exists as both a liquid and a vapor simultaneously, with distinct properties? Understanding critical points is essential for applications ranging from industrial processes to environmental studies.

6.4. THERMODYNAMIC EQUILIBRIUM

Thermodynamic equilibrium refers to a state in which all macroscopic variables that characterize a system remain constant over time. These variables include temperature, pressure, volume, and chemical composition. At equilibrium, there is no net transfer of energy or matter within the system or between the system and its surroundings (Extrand, 2022). This condition is governed by the second law of thermodynamics, which states that natural processes tend toward increasing entropy and achieving a state of maximum disorder. Understanding thermodynamic equilibrium is essential in fields such as chemistry, physics, and engineering, where it underpins the analysis of energy transformations, chemical reactions, and the behavior of complex systems ranging from molecular interactions to planetary atmospheres.

6.4.1. Gibbs Phase Rule

The Gibbs phase rule, named after Josiah Willard Gibbs, is a fundamental equation used to predict the number of variables that must be specified to define the equilibrium state of a system in different phases and conditions.

The Gibbs phase rule is expressed as:

$$F = C - P + 2$$

where; (F) is the degrees of freedom (the number of intensive variables that can be independently varied without changing the number of phases in equilibrium); (C) is the number of components (chemically independent constituents); and (P) is the number of phases in equilibrium. This rule helps in understanding the relationships between the number of components, phases, and variables in a system at equilibrium. For example, in a simple single-phase system with no chemical reactions, the degrees of freedom (F) would be 2 (temperature and pressure), indicating that these two variables can be varied independently while keeping the system in a single phase equilibrium (Egorov et al., 2003).

6.4.2. Chemical Potential and Equilibrium Conditions

Chemical potential (μ) is a fundamental concept in thermodynamics that describes the change in Gibbs free $energy(G)$ of a system as a function of the number of particles

(N), temperature (T), and pressure (P). It is defined as:

$$\mu_i = \left(\frac{\partial G}{\partial N_i}\right)_{T,P,N_j}$$

where; (N_i) is the number of particles of component (i); and ( ) represents all other components.

The chemical potential determines the direction of spontaneous processes and phase equilibrium conditions. At equilibrium, the chemical potential of each component must be equal in all phases in contact. This principle is expressed by the Gibbs-Duhem equation:

$$\sum_i N_i \, d\mu_i = 0$$

This equation ensures that changes in chemical potentials across different phases are balanced, maintaining equilibrium conditions.

6.4.3. Phase Equilibria in Mixtures

Phase equilibria in mixtures involve the distribution of components between different phases at equilibrium. Key concepts include:

For mixtures, the phase rule is modified to account for the distribution of components between phases:

$$F = C - P + 2$$

where; (C) is the number of components; (P) is the number of phases; and ( ) is the degrees of freedom.

Partial molar quantities $\left(\overline{X_i}\right)$ describe the contribution of component (i) to the total property of the mixture (e.g., partial molar volume $\left(\overline{V_i}\right)$. They are defined as:

$$\overline{X_i} = \left(\frac{\partial(nX)}{\partial n_i}\right)_{T,P,n_j}$$

Where; (X) represents the property of interest (e.g., volume, entropy).

Phase diagrams for mixtures illustrate the composition and phase behavior as a function of temperature and pressure. These diagrams show regions of single-phase and multiphase equilibrium, along with critical points and phase boundaries (Andrukh et al., 2014).

6.4.4. Applications to Atmospheric Physics and Climate

In atmospheric physics, thermodynamic equilibrium principles are essential for understanding the distribution of gases, aerosols, and energy exchanges in the atmosphere. Chemical equilibrium between different atmospheric components (e.g., gases, water vapor) influences atmospheric composition and reactions.

Climate equilibrium involves the balance between incoming solar radiation and outgoing infrared radiation, determining Earth's temperature and climate stability. Radiative equilibrium models consider factors such as albedo, greenhouse gases, and cloud cover to predict global temperature trends.

Climate models utilize thermodynamic principles to simulate energy exchanges, cloud formation, and atmospheric circulation. Equilibrium states are used to analyze feedback mechanisms and predict long-term climate trends under different scenarios (e.g., greenhouse gas emissions). In conclusion, the concepts of Gibbs phase rule, chemical potential, phase equilibria in mixtures, and their applications in atmospheric physics and climate science provide a foundational understanding of equilibrium conditions and phase behavior in complex systems. These principles are crucial for advancing scientific knowledge, modeling natural phenomena, and addressing environmental challenges related to climate change and atmospheric dynamics.

ACTIVITY 6.1.

Explore capillary action and meniscus formation in liquids. Conduct experiments using different liquids and materials to observe how surface tension and adhesive forces contribute to phenomena like capillary rise and meniscus curvature.

SUMMARY

- Chapter Six delves into the intricate properties and behaviors of liquids and gases, providing a comprehensive analysis of their molecular structures and interactions. It begins by examining the fundamental properties of liquids, such as density and compressibility, and explores more complex characteristics like viscosity and surface tension. The phenomena of capillary action and meniscus formation are also elucidated, highlighting the significance of molecular interactions.
- In discussing gases, the chapter introduces the ideal gas law, explaining its foundational principles and the limitations that lead to deviations in real gases. The kinetic theory of gases is presented as a framework for understanding gas behavior, and the chapter explores the equations of state that account for deviations in real gases. Additionally, gas mixtures and the concept of partial pressures are addressed, providing a thorough understanding of gas dynamics.
- The chapter further investigates phase transitions, detailing processes such as vaporization, condensation, melting, freezing, sublimation, and deposition. The concept of critical points and the interpretation of phase diagrams are explained, emphasizing the thermodynamic principles governing these transitions. The discussion on thermodynamic equilibria introduces the Gibbs phase rule and the concept of chemical potential, applying these principles to phase equilibria in mixtures.
- The chapter concludes by connecting these thermodynamic concepts to real-world applications, particularly in atmospheric physics and climate science. The exploration of how phase equilibria and thermodynamic principles influence atmospheric phenomena and climate dynamics underscores the practical relevance of the theoretical concepts discussed.

REVIEW QUESTIONS

1. What are the key molecular interactions in liquids, and how do they affect properties like viscosity and surface tension?
2. How does the ideal gas law describe the behavior of gases, and what are the limitations of this law?
3. Explain the kinetic theory of gases and how it relates to the properties of gases.
4. Describe the deviations observed in real gases and the equations of state used to account for these deviations.
5. What are the different types of phase transitions, and what thermodynamic principles govern these transitions?
6. How is the Gibbs phase rule used to determine the number of phases in equilibrium?
7. Discuss the application of phase equilibria concepts to atmospheric physics and climate.

MULTIPLE CHOICE QUESTIONS

1. **What property of liquids is primarily influenced by molecular interactions?**
   a. compressibility
   b. viscosity
   c. density
   d. partial pressure
2. **Which law describes the relationship between the pressure, volume, and temperature of an ideal gas?**
   a. Boyle's Law
   b. Charles's Law
   c. Ideal Gas Law
   d. Avogadro's Law
3. **The kinetic theory of gases is based on which assumption about gas particles?**
   a. They have significant volume.
   b. They are in constant random motion.
   c. They interact with strong intermolecular forces.
   d. They lose energy during collisions.
4. **Which phase transition involves a substance changing directly from a solid to a gas?**
   a. melting
   b. condensation
   c. sublimation
   d. deposition
5. **The Gibbs phase rule is used to determine:**
   a. the density of a liquid
   b. the behavior of real gases
   c. the number of phases in equilibrium
   d. the viscosity of a liquid

Answers to Multiple Questions

1. (b); 2. (c); 3. (b); 4. (c); 5. (c).

REFERENCES

1. Andrukh, T., Monaenkova, D., Rubin, B., Lee, W. K., & Kornev, K. G. (2014). Meniscus formation in a capillary and the role of contact line friction. *Soft Matter, 10*(4), 609–615.

2. Brooks, R. F., Egry, I., Seetharaman, S., & Grant, D. (2001). Reliable data for high-temperature viscosity and surface tension: Results from a European project. *High Temperatures-High Pressures, 33*(6), 631–637.

3. Burbury, S. H. (1894). On the kinetic theory of gases. *Science Progress (1894–1898), 2*(9), 203–224.

4. Butt, H. J., & Kappl, M. (2009). Normal capillary forces. *Advances in Colloid and Interface Science, 146*(1–2), 48–60.

5. Carnahan, N. F., & Starling, K. E. (1972). Intermolecular repulsions and the equation of state for fluids. *AIChE Journal, 18*(6), 1184–1189.

6. Chothia, C., & Jones, E. Y. (1997). The molecular structure of cell adhesion molecules. *Annual Review of Biochemistry, 66*(1), 823–862.

7. Cistola, D. P., Hamilton, J. A., Jackson, D., & Small, D. M. (1988). Ionization and phase behavior of fatty acids in water: Application of the Gibbs phase rule. *Biochemistry, 27*(6), 1881–1888.

8. Domańska, U. (2006). Thermophysical properties and thermodynamic phase behavior of ionic liquids. *Thermochimica Acta, 448*(1), 19–30.

9. Egorov, A. G., Kornev, K. G., & Neimark, A. V. (2003). Meniscus motion in a prewetted capillary. *Physics of Fluids, 15*(10), 3134–3143.

10. Ehlers, J. (2005). Kinetic theory of gases in general relativity theory. In *Lectures in Statistical Physics: Advanced School for Statistical Mechanics and Thermodynamics Austin, Texas/USA* (pp. 78–105).

11. Ellis, R. S., & Ellis, R. S. (1985). Large deviations and the discrete ideal gas. In *Entropy, Large Deviations, and Statistical Mechanics* (pp. 59–87).

12. Elsharkawy, A. M. (2004). Efficient methods for calculations of compressibility, density, and viscosity of natural gases. *Fluid Phase Equilibria, 218*(1), 1–13.

13. Extrand, C. W. (2022). Meniscus formation in a vertical capillary tube. *Langmuir, 38*(7), 2346–2353.

14. Fleischer, C. C., & Payne, C. K. (2014). Nanoparticle–Cell interactions: Molecular structure of the protein corona and cellular outcomes. *Accounts of Chemical Research, 47*(8), 2651–2659.

15. Fowler, M. (2008). Kinetic theory of gases: A brief review. *Physics of Fluids, 15*(2), 1–13.

16. Fredlund, D. G. (1976). Density and compressibility characteristics of air–Water mixtures. *Canadian Geotechnical Journal, 13*(4), 386–396.

17. Gaiduk, A. P., Gygi, F., & Galli, G. (2015). Density and compressibility of liquid water and ice from first-principles simulations with hybrid functionals. *The Journal of Physical Chemistry Letters, 6*(15), 2902–2908.

18. Gupta, S. V. (2004). Capillary action in narrow and wide tubes—A unified approach. *Metrologia, 41*(6), 361.

19. Hossain, A. K., & Davies, P. A. (2013). Pyrolysis liquids and gases as alternative fuels in internal combustion engines–A review. *Renewable and Sustainable Energy Reviews, 21*, 165–189.

20. Jaeger, H. M., Nagel, S. R., & Behringer, R. P. (1996). Granular solids, liquids, and gases. *Reviews of Modern Physics, 68*(4), 1259.
21. Johansson, J., & Curstedt, T. (1997). Molecular structures and interactions of pulmonary surfactant components. *European Journal of Biochemistry, 244*(3), 675–693.
22. Kauzmann, W. (1948). The nature of the glassy state and the behavior of liquids at low temperatures. *Chemical Reviews, 43*(2), 219–256.
23. Khattab, I. S., Bandarkar, F., Fakhree, M. A. A., & Jouyban, A. (2012). Density, viscosity, and surface tension of water+ ethanol mixtures from 293 to 323K. *Korean Journal of Chemical Engineering, 29*, 812–817.
24. Kilian, H. G. (1981). Equation of state of real networks. *Polymer, 22*(2), 209–217.
25. Kim, K. S., Tarakeshwar, P., & Lee, J. Y. (2000). Molecular clusters of π-systems: Theoretical studies of structures, spectra, and origin of interaction energies. *Chemical Reviews, 100*(11), 4145–4186.
26. Landau, L. (1936). The theory of phase transitions. *Nature, 138*(3498), 840–841.
27. Lehtonen, T., & Nummelin, E. (1990). Level I theory of large deviations in the ideal gas. *International Journal of Theoretical Physics, 29*, 621–635.
28. Liu, X., Godbole, A., Lu, C., Michal, G., & Venton, P. (2014). Source strength and dispersion of CO_2 releases from high-pressure pipelines: CFD model using real gas equation of state. *Applied Energy, 126*, 56–68.
29. Loitsyanskiy, L. G. (1966). Mechanics of liquids and gases. *Prikladnaya Mekhanika, 6*, 17–18.
30. Mao, H. K., Chen, X. J., Ding, Y., Li, B., & Wang, L. (2018). Solids, liquids, and gases under high pressure. *Reviews of Modern Physics, 90*(1), 015007.
31. Norton, S. J. (1983). Generation of separate density and compressibility images in tissue. *Ultrasonic Imaging, 5*(3), 240–252.
32. Pelofsky, A. H. (1966). Surface tension-viscosity relation for liquids. *Journal of Chemical and Engineering Data, 11*(3), 394–397.
33. Ranga Raju, K. G., & Asawa, G. L. (1977). Viscosity and surface tension effects on weir flow. *Journal of the Hydraulics Division, 103*(10), 1227–1231.
34. Rebelo, L. P. N., Najdanovic-Visak, V., Gomes de Azevedo, R., Esperança, J. M. S. S., Nunes da Ponte, M., Guedes, H. J. R., & Cordeiro, T. C. (2005). Phase behavior and thermodynamic properties of ionic liquids, ionic liquid mixtures, and ionic liquid solutions. In *Ionic Liquids IIIA: Fundamentals, Progress, Challenges, and Opportunities—Properties and Structure* (Vol. 901, pp. 270–291).
35. Rooney, D., Jacquemin, J., & Gardas, R. (2010). Thermophysical properties of ionic liquids. In *Ionic Liquids* (pp. 185–212).
36. Rumble III, D. (1974). Gibbs phase rule and its application in geochemistry. *Journal of the Washington Academy of Sciences, 64*(7), 199–208.
37. Sahasrabudhe, S. N., Rodriguez-Martinez, V., O'Meara, M., & Farkas, B. E. (2017). Density, viscosity, and surface tension of five vegetable oils at elevated temperatures: Measurement and modeling. *International Journal of Food Properties, 20*(sup2), 1965–1981.

38. Sánchez, A., & Villalobos, S. (2023). Geometric model of real gases described by the Redlich-Kwong-Soave equation of state. *Journal of Molecular Liquids, 372*, 121–167.

39. Stepanov, I. (2017). Exact calculation of the internal energy of the ideal gas in statistical mechanics. *Physical Science International Journal, 14*(2), 1–5.

40. Sun, W., & Powell-Palm, M. J. (2021). Generalized Gibbs' phase rule. *arXiv preprint arXiv:2105.01337*.

41. Taylor, C. D. (1978). The effect of pressure upon the solubility of oxygen in water: Implications of the deviation from the ideal gas law upon measurements of fluorescence quenching. *Archives of Biochemistry and Biophysics, 191*(1), 375–384.

42. Travis, J. R., Koch, D. P., Xiao, J., & Xu, Z. (2013). Real-gas equations-of-state for the GASFLOW CFD code. *International Journal of Hydrogen Energy, 38*(19), 8132–8140.

43. Uhlenbeck, G. E., & Beth, E. (1936). The quantum theory of the non-ideal gas I. Deviations from the classical theory. *Physica, 3*(8), 729–745.

44. Vovchenko, V. (2017). Equations of state for real gases on the nuclear scale. *Physical Review C, 96*(1), 015–020.

CHAPTER

7 Soft Matter and Complex Fluids

LEARNING OBJECTIVES

After studying this chapter, you will be able to:

1. Understand the structural basis and phase behavior of soft matter;
2. Comprehend the dynamic nature and properties of soft matter and its applications in science and technology;
3. Identify and describe the different types of polymers, colloids, and liquid crystals;
4. Analyze the viscoelastic behavior and shear and flow characteristics of complex fluids;
5. Explore the engineering and biomedical applications of soft matter and complex fluids;
6. Examine the phenomena of emergent properties and self-assembly in soft matter systems;
7. Investigate the biological and medical applications of soft matter and complex fluids.

KEY TERMS FROM THIS CHAPTER

Colloids
Liquid Crystals
Self-Assembly
Soft Matter
Emergent Properties
Polymers
Shear and Flow Characteristics
Viscoelastic Behavior

UNIT INTRODUCTION

Soft matter and complex fluids are intriguing areas of materials science that exist between the conventional states of solids and liquids. These materials exhibit distinct properties, often characterized by their ability to flow and deform under applied stresses while maintaining some degree of structural organization (Fuller & Vermant, 2011).

Soft Matter includes a wide array of materials such as polymers, colloids, liquid crystals, and biological macromolecules. These materials are generally flexible, responsive to external stimuli, and capable of self-assembling into ordered structures. Soft matter is integral to many scientific disciplines and technologies, including biomedical applications and advanced materials engineering (Chaudhuri et al., 2011).

Complex Fluids are fluids that display non-Newtonian behavior, where their viscosity and flow properties change under different conditions. This category includes suspensions, emulsions, and gels. The behavior of these fluids often depends on factors such as particle size, concentration, and external forces, making them highly applicable in industries like food science, cosmetics, and pharmaceuticals (Nagel, 2017).

Understanding soft matter and complex fluids involves studying their molecular structures, phase behaviors, rheological properties, and applications across various industries and research fields. This area of study is continually evolving, offering new insights into the fundamental principles of material behavior and driving innovations in technology and medicine (Hatzikiriakos, 2015).

7.1. SOFT MATTER

Soft matter comprises a diverse class of materials distinguished by their intermediate properties between conventional solids and liquids. These substances are characterized by their ability to undergo large deformations under external forces while maintaining their structural integrity. Soft matter includes polymers, colloids, liquid crystals, and biological materials, each exhibiting unique behaviors such as self-assembly and responsiveness to environmental changes (Brown, 2017). Understanding soft matter is crucial for advancing fields like materials science, biophysics, and nanotechnology, offering insights into complex phenomena such as phase transitions, mechanical properties, and the design of novel materials for various applications (Tabor et al., 2011) (Figure 7.1).

Figure 7.1. *Biological and physical sciences' soft matter highlighting its: colloids granular media, foams, polymers, non-Newtonian fluids, rheology, and active matter research.*

Source: NASA Glenn Research Center, Creative Commons License.

7.1.1. Structural Basis

Soft matter systems are characterized by their hierarchical structures, which range from molecular to macroscopic scales. At the molecular level, interactions such as van der Waals forces, hydrogen bonding, and electrostatic interactions play significant roles in determining the material's properties. These interactions can be quantitatively described using potential energy functions, such as the Lennard-Jones potential for van der Waals forces:

$$V_{LJ}(r) = 4\epsilon\left[\left(\frac{\sigma}{r}\right)^{12} - \left(\frac{\sigma}{r}\right)^{6}\right]$$

where; (ϵ) is the depth of the potential well; (σ) is the finite distance at which the inter-particle potential is zero; and (r) is the distance between the particles (Tanaka & Araki, 2006).

Remember

Remember that soft matter encompasses a diverse range of materials with unique structural and dynamic properties, often exhibiting complex phase behavior and applications across various scientific and technological fields.

7.1.2. Phase Behavior

Soft matter materials can exist in various phases (solid, liquid, gas, and more complex phases like liquid crystals), depending on external conditions such as temperature, pressure, and chemical composition. Phase transitions between these states are critical for understanding the material's response to stimuli and its applications (Huber, 2015).

7.1.3. Dynamic Nature and Properties

Soft matter exhibits dynamic behavior due to the continuous rearrangement of its constituent molecules or particles. This dynamic nature underlies several important properties that distinguish soft matter from conventional materials.

7.1.3.1. Viscoelasticity

Soft matter materials typically display viscoelastic behavior, characterized by both viscous (flow-like) and elastic (reversible deformation) responses under stress. This behavior is described by viscoelastic models such as the Maxwell model, which relates stress (σ), strain (ϵ), and relaxation time (τ):

$$\sigma(t) = E\epsilon(t) + \eta\frac{d\epsilon(t)}{dt}$$

where; (E) is the elastic modulus; and (η) is the viscosity (Roberto & Véronique, 2023).

7.1.3.2. Brownian Motion

In colloidal systems, Brownian motion describes the random movement of particles suspended in a fluid due to collisions with surrounding molecules. This phenomenon is modeled using the Langevin equation:

$$m\frac{d^2r}{dt^2} = -\nabla U(r) + \sqrt{2k_BT\gamma}\xi(t)$$

where; (m) is the particle mass; $U(r)$ is the potential energy; (k_B) is Boltzmann's constant; (T) is temperature; $(\tilde{a})$ is the friction coefficient; and $(\hat{\imath}(t))$ is a Gaussian white noise term (Haw, 2002).

7.1.4. Applications in Science and Technology

Soft matter finds extensive applications across various disciplines, owing to its unique properties and responsiveness to external stimuli.

1. **Biomedical Applications:** Soft matter materials are used in drug delivery systems, tissue engineering scaffolds, and medical diagnostics due to their biocompatibility and ability to mimic biological environments (Lagerwall & Scalia, 2012).
2. **Engineering and Materials Science:** In engineering, soft matter is utilized in designing responsive materials, actuators, and sensors. Liquid crystal displays (LCDs) are a notable example where the orientation of liquid crystal molecules changes in response to an electric field, enabling display technology.
3. **Environmental and Energy Applications:** Soft matter materials are investigated for applications in energy storage (e.g., battery electrolytes) and environmental remediation (e.g., wastewater treatment), leveraging their ability to interact with and encapsulate molecules and nanoparticles.

This comprehensive understanding of fundamental concepts, dynamic properties, and applications underscores the importance of soft matter in advancing scientific knowledge and technological innovations across various fields (Grosjean et al., 2019).

Did you know?

Did you know that soft matter materials find extensive applications in science and technology, ranging from drug delivery systems in medicine to advanced materials in electronics? These applications highlight the versatility and importance of understanding soft matter behavior.

7.2. POLYMERS, COLLOIDS, AND LIQUID CRYSTALS

Polymers, colloids, and liquid crystals are key categories within soft matter, each contributing distinct characteristics to materials science and technology (Nakayama & Kato, 2022).

7.2.1. Polymers

7.2.1.1. Molecular Structures and Behavior

Polymers are macromolecules consisting of repeated units called monomers, linked together through covalent bonds. The molecular structure of polymers greatly influences their properties and behavior. The arrangement of monomers and the type of bonding (e.g., linear, branched, cross-linked) dictate characteristics such as flexibility, strength, and chemical resistance. For instance, the tacticity of polymers—how monomers are arranged in space—affects their crystallinity and mechanical properties. The behavior of polymers in solution or in the solid state is often described using polymer solution theory and principles of polymer physics, which involve statistical mechanics to analyze the distribution of molecular weights and chain conformations (Kim et al., 2019) (Figure 7.2).

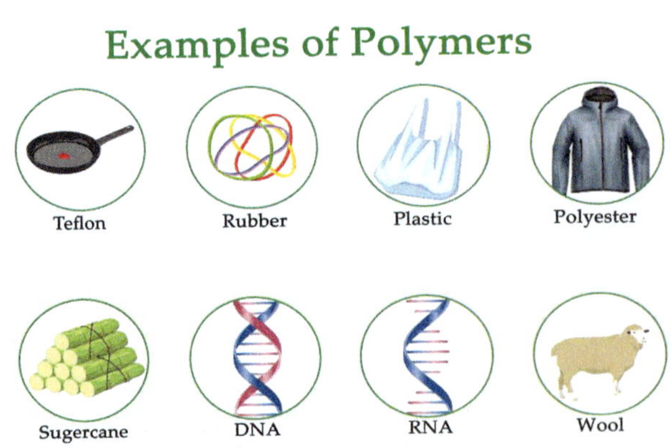

Figure 7.2. *Examples of polymers.*

Source: Geeks for Geeks, Creative Commons License.

7.2.1.2. Polymerization and Chain Conformations

Polymerization is the process by which monomers react to form polymers. This process can be initiated by various methods, such as radical polymerization, where free radicals induce chain growth by adding monomers successively. The growth and structure of polymer chains depend on reaction conditions like temperature, solvent, and catalysts. Chain conformations, or the spatial arrangements of polymer chains, play a critical role in determining polymer properties. Models like the random coil model describe polymers as flexible chains that assume random configurations in solution, while more ordered structures, such as the extended chain configuration, influence properties like tensile strength and elasticity. Theoretical frameworks like the Flory–Huggins theory incorporate polymer-solvent interactions and entanglement effects, crucial for understanding polymer behavior in solution and polymer blends (Mark, 1979).

Tip

Polymers are large molecules composed of repeating structural units (monomers) connected by covalent bonds. Understanding polymer structure and properties is crucial for designing materials with specific mechanical, thermal, and chemical characteristics.

7.2.1.3. Mechanical Properties

The mechanical properties of polymers are defined by their response to external forces, crucial for applications ranging from industrial materials to biomedical devices. Key mechanical properties include:

1. **Elastic Modulus (E):** Measures the stiffness of a polymer and describes its ability to return to its original shape after deformation. It is quantified using stress-strain curves, where stress (force per unit area) is plotted against strain (relative deformation).

   $$E = \frac{\sigma}{\epsilon}$$

   where; (σ) is stress; and (ϵ) is strain.
2. **Tensile Strength:** Maximum stress a polymer can withstand before breaking, indicative of its durability and load-bearing capacity (Daniels et al., 1990).
3. **Viscoelasticity:** Polymer chains exhibit both viscous (flow-like) and elastic (reversible deformation) behavior under stress, characterized by relaxation processes and time-dependent responses.

Understanding these properties requires mathematical models such as the Maxwell model or the Voigt model, which describe viscoelastic behavior by combining elements of elastic and viscous responses in polymers. Overall, the molecular structure, polymerization methods, and mechanical properties of polymers are intricately linked, defining their utility across diverse technological applications (Nunes et al., 1982).

7.2.2. Colloids

7.2.2.1. Characteristics and Stability

Colloids are heterogeneous systems where one substance (the dispersed phase) is finely divided or distributed within another substance (the continuous phase), typically a liquid or gas. The dispersed phase consists of particles ranging in size from nanometers to micrometers, often too small to settle under gravity but large enough to scatter light, resulting in phenomena like Tyndall effect. Colloids can be categorized based on the dispersed phase's state (solid, liquid, or gas) and the dispersion medium's phase. Stability of colloidal systems is influenced by factors such as electrostatic repulsion, steric hindrance, and Van der Waals forces. These factors are quantitatively described by DLVO theory, which considers the balance of attractive and repulsive forces between colloidal particles in a medium (Liu et al., 2018) (Figure 7.3).

Figure 7.3. *Examples of colloids.*

Source: Geeks for Geeks, Creative Commons License.

7.2.2.2. Brownian Motion and Dynamics

In colloidal suspensions, particles undergo continuous, random motion known as Brownian motion. This movement is caused by collisions with solvent molecules, influencing the dispersion's dynamics and properties. The behavior of colloidal particles in suspension is mathematically described using the Langevin equation:

$$m\frac{d^2r}{dt^2} = -\nabla U(r) + \xi(t)$$

where; (m) is the particle mass; (r) is the position vector; $U(r)$ is the potential energy; and $\xi(t)$ is a random force term representing thermal fluctuations.

7.2.2.3. Industrial Applications

Colloids play pivotal roles in various industrial applications due to their unique properties and behaviors. They are extensively used in:

1. **Cosmetics and Pharmaceuticals:** Colloidal dispersions stabilize active ingredients, improve product appearance, and enable controlled drug delivery systems.
2. **Food Science:** Colloids enhance texture, stability, and mouthfeel in food products like emulsions, suspensions, and foams (Khan & Khan, 2014).
3. **Materials Science:** Colloidal suspensions are employed in the manufacture of ceramics, catalysts, and coatings, where particle size and distribution influence material properties and performance.
4. **Environmental Engineering:** Colloids are utilized in water treatment processes to remove contaminants through mechanisms like flocculation and coagulation.

The study of colloids involves advanced techniques such as dynamic light scattering (DLS) and electrophoresis to characterize particle size, charge, and stability. Understanding colloidal dynamics and stability is essential for optimizing their industrial applications, ensuring efficiency, and enhancing product performance across diverse sectors (Aguirre et al., 2023).

7.2.3. Liquid Crystals

7.2.3.1. Types and Molecular Alignment

Liquid crystals (LCs) are substances that exhibit properties between those of conventional liquids and solid crystals. They are characterized by anisotropic molecular arrangements, where the orientation of molecules exhibits order in one or more dimensions while retaining fluid-like mobility in others. Depending on molecular alignment, liquid crystals can be classified into several types (Figure 7.4):

1. **Nematic:** Molecules align along a preferred direction (director), with no positional order.
2. **Smectic:** Molecules arrange in layers, each layer exhibiting positional order while maintaining orientational order within layers.

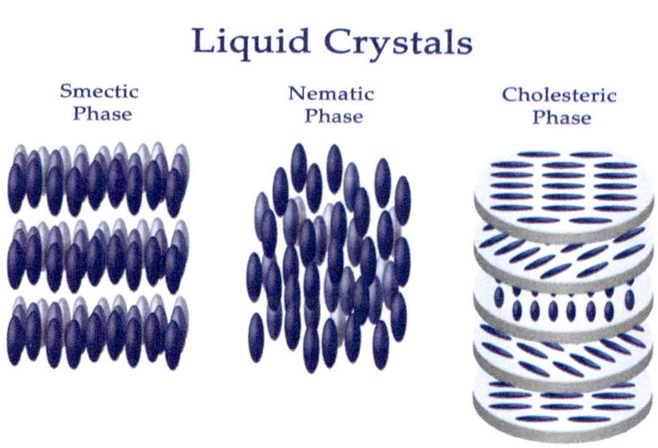

Figure 7.4. *Classification of liquid crystals.*

Source: Smart Glass World, Creative Commons License.

3. **Cholesteric (Chiral Nematic):** Molecules arrange in layers with a helical structure, resulting in wavelength-dependent reflection properties.

The alignment of liquid crystal molecules is influenced by external factors such as temperature, electric fields, and surface treatments, which control the orientation and phase transitions between different LC phases (Liu & Broer, 2014).

7.2.3.2. Phases and Optical Properties

Liquid crystals exhibit distinct phases (nematic, smectic, cholesteric) depending on temperature and molecular structure. Each phase transition involves changes in molecular order and optical properties. For example, in nematic liquid crystals, the director's orientation can be manipulated using electric fields (electro-optic effect), affecting light polarization and transmission. Optical properties of liquid crystals are described using the refractive index tensor (n_{ij}), which varies with direction due to molecular alignment. The birefringence (Än), a measure of the difference between refractive indices along different axes, determines LCs' ability to manipulate light polarization in display technologies.

7.2.3.3. Display and Optoelectronic Applications

Liquid crystals play a crucial role in display technologies, particularly in liquid crystal displays (LCDs). In LCDs, thin layers of nematic liquid crystals are sandwiched between transparent electrodes. By applying voltage across these electrodes, the orientation of liquid crystal molecules changes, altering light polarization and transmission through the display. This principle, known as the electrically controlled birefringence, enables the creation of pixels that selectively block or transmit light, forming images visible to the viewer. The operational principle involves the Frank–Oseen elasticity theory, describing the energy associated with deformations in liquid crystal director fields under external influences.

Beyond displays, liquid crystals find applications in optoelectronic devices such as variable optical attenuators (VOAs), light modulators, and polarization controllers. The ability to control light polarization and transmission makes liquid crystals indispensable in telecommunications, imaging systems, and optical sensors.

Understanding the types, phases, and optical properties of liquid crystals is essential for optimizing their applications in display technologies and advancing their utility in emerging fields of optoelectronics and photonics (Lin et al., 2015).

7.3. RHEOLOGY AND MECHANICAL PROPERTIES

Rheology is the study of how materials deform and flow under applied stress, encompassing their mechanical properties in response to forces and strains. In the context of soft matter like polymers, colloids, and liquid crystals, rheology plays a crucial role in understanding and predicting their behavior in various applications. It involves measuring parameters such as viscosity, elasticity, and viscoelasticity, which characterize how these materials respond to different deformation rates and temperatures. Rheological studies utilize mathematical models like the Maxwell model and the Kelvin-Voigt model to describe viscoelastic behavior, crucial for optimizing processing conditions in industries ranging from food and cosmetics to pharmaceuticals and materials science. Understanding rheology helps engineers and scientists tailor the mechanical properties of soft matter to meet specific performance criteria, enhancing product design and manufacturing efficiency (Wu et al., 2019).

7.3.1. Viscoelastic Behavior

Viscoelastic behavior describes materials that exhibit both viscous (flow-like) and elastic (reversible deformation) properties when subjected to stress. This behavior is characterized by a time-dependent response to applied forces, where the material's deformation rate varies over time. The stress-strain relationship in viscoelastic materials can be described using the generalized form of Hooke's law:

$$\sigma_{ij}(t) = \sum_{k,l} C_{ijkl}\left(\epsilon_{kl}(t) - \epsilon_{kl}^{0}\right)$$

where; $\sigma_{ij}(t)$ is the stress tensor; $\epsilon_{kl}(t)$ is the strain tensor; (C_{ijkl}) are the components of the fourth-order elasticity tensor; and $\left(\epsilon_{kl}^{0}\right)$ represents the initial strain.

Viscoelastic materials exhibit phenomena such as creep (continuous deformation under constant stress) and stress relaxation (gradual decrease in stress under constant strain), which are modeled using viscoelastic constitutive equations like the Maxwell model and the Kelvin-Voigt model. These models incorporate elements of spring (elastic) and dashpot (viscous) to describe the material's response under different loading conditions.

7.3.2. Shear and Flow Characteristics

The shear and flow characteristics of materials describe how they deform and flow under applied shear stress. In rheology, shear stress (τ) is defined as the force per unit area acting parallel to a plane within the material, while shear rate $(\dot{\gamma})$ is the rate of change of deformation per unit time. The relationship between shear stress and shear rate is described by the constitutive equation:

$$\tau = \eta \cdot \dot{\gamma}$$

where; (η) is the viscosity, representing the material's resistance to flow. For viscoelastic materials, viscosity (η)varies with shear rate and time, indicating their complex response to deformation.

Flow characteristics are influenced by factors such as temperature, pressure, and molecular structure. Non-Newtonian fluids, including viscoelastic materials, exhibit shear-thinning (decreasing viscosity with increasing shear rate) or shear-thickening (increasing viscosity with increasing shear rate) behavior. The flow behavior index (n) quantifies this relationship in power-law fluids (Richards et al., 2021):

$$\tau = K \cdot (\dot{\gamma})^n$$

where; (K) is the consistency index.

7.3.3. Engineering and Biomedical Applications

Viscoelastic materials find diverse applications in engineering and biomedical fields due to their tailored mechanical properties and responsiveness under stress. In engineering, these materials are used in damping applications to absorb mechanical vibrations and shocks, enhancing structural stability and durability. Rheological characterization helps engineers design materials with specific viscoelastic properties to meet performance requirements in automotive components, aerospace structures, and damping devices.

In biomedical applications, viscoelastic materials are crucial for soft tissue engineering, prosthetics, and drug delivery systems. Biocompatible polymers with controlled viscoelastic behavior mimic natural tissues' mechanical properties, promoting tissue regeneration and enhancing patient comfort. Rheological studies guide the development of injectable hydrogels and scaffolds that support cell growth and tissue repair.

Overall, understanding the viscoelastic behavior, shear and flow characteristics, and their applications is essential for optimizing material design and performance across engineering and biomedical disciplines, ensuring efficient and innovative technological advancements (Chen et al., 2020).

7.4. APPLICATIONS AND PHENOMENA

The applications are explained in the following sections.

7.4.1. Biological and Medical Applications

Liquid crystals, polymers, colloids, and other soft matter exhibit unique properties that make them valuable in biological and medical applications. In drug delivery systems, for instance, colloidal suspensions can encapsulate drugs and release them in a controlled manner, ensuring targeted delivery and minimizing side effects. The self-assembly properties of polymers and liquid crystals enable the creation of biomimetic materials that mimic natural tissues' mechanical and biological functions. Mathematical models such as diffusion equations and pharmacokinetic models are employed to predict drug release kinetics and optimize therapeutic efficacy. Moreover, liquid crystal-based biosensors leverage the sensitivity of LC phases to detect biomolecular interactions, offering rapid and sensitive diagnostic tools in medical diagnostics (Christov, 2021).

7.4.2. Technological Innovations

Soft matter plays a pivotal role in technological advancements across various industries. Liquid crystal displays (LCDs) are prime examples where the alignment of liquid crystal molecules under applied electric fields modulates light transmission, enabling vibrant display technologies. The optical properties of liquid crystals are mathematically modeled using refractive indices and polarization matrices to optimize display performance. In nanotechnology, colloidal suspensions are engineered with specific size and surface properties to enhance catalytic efficiency in fuel cells and environmental remediation. These innovations rely on mathematical formulations such as surface area calculations and reaction rate equations to design and scale-up manufacturing processes effectively (Saraswat & Sengwa, 2024).

7.4.3. Emergent Properties and Self-Assembly

Emergent properties in soft matter arise from the collective interactions of constituent molecules, leading to novel behaviors not exhibited by individual components. Self-assembly processes, governed by thermodynamic principles and statistical mechanics, drive the formation of complex structures such as colloidal crystals and liquid crystal domains. Mathematical models like Gibbs free energy minimization and Monte Carlo simulations elucidate the equilibrium configurations and phase transitions in these systems. Understanding emergent properties facilitates the design of functional materials with tailored mechanical, optical, and electronic properties. Applications range from photonics and flexible electronics to adaptive materials that respond to environmental stimuli. In conclusion, the diverse applications and phenomena of soft matter underscore its interdisciplinary significance in biological, medical, technological, and fundamental research domains. Mathematical formulations and theoretical frameworks play a crucial role in elucidating and harnessing the unique properties of soft matter for innovative applications and scientific advancements (Stupp et al., 2013).

ACTIVITY 7.1.

Explore the shear and flow characteristics of soft matter materials through rheological experiments. Measure viscosity and viscoelastic properties to understand how these materials respond to mechanical forces, critical for engineering and biomedical applications.

SUMMARY

- Chapter Seven explores the fascinating realm of soft matter and complex fluids, focusing on their structural, dynamic, and application-oriented aspects. The chapter begins with an introduction to the structural basis of soft matter, highlighting its unique phase behavior and dynamic nature. This is followed by a detailed examination of polymers, colloids, and liquid crystals, discussing their properties, behaviors, and roles in various scientific and technological contexts.
- The chapter delves into the rheological and mechanical properties of these materials, particularly viscoelastic behavior and shear and flow characteristics, which are crucial for understanding their functionality in different applications. It also emphasizes the engineering and biomedical applications of soft matter and complex fluids, showcasing their importance in innovation and practical use.
- Additionally, the chapter covers the phenomena of emergent properties and self-assembly, providing insights into the complex behaviors and interactions within soft matter systems. The biological and medical applications of these materials are also discussed, illustrating their relevance in modern science and technology.

REVIEW QUESTIONS

1. Explain the structural basis of soft matter and its phase behavior.
2. Describe the different types of polymers and their properties.
3. What are colloids and how do they differ from other complex fluids?
4. Discuss the characteristics of liquid crystals and their applications.
5. Explain viscoelastic behavior and its significance in the study of soft matter.
6. What are some of the technological innovations that have arisen from the study of soft matter and complex fluids?

MULTIPLE CHOICE QUESTIONS

1. Which of the following is a characteristic property of soft matter?

a. High rigidity

b. High viscosity

c. High elasticity

d. High density

2. Polymers are composed of:

a. Small molecules

b. Ions

c. Macromolecules

d. Colloidal particles

3. **Liquid crystals are used in:**
   a. Optical fibers
   b. Television screens
   c. Catalysts
   d. Batteries
4. **Viscoelastic behavior is a property of:**
   a. Metals
   b. Liquids
   c. Gases
   d. Soft matter
5. **Self-assembly in soft matter systems results in:**
   a. Disordered structures
   b. Crystalline solids
   c. Emergent properties
   d. Static behavior

Answers to Multiple Questions

1. (c); 2. (c); 3. (d); 4. (b); 5. (c).

REFERENCES

1. Aguirre, M., Ballard, N., Gonzalez, E., Hamzehlou, S., Sardon, H., Calderon, M., & Asua, J. M. (2023). Polymer colloids: Current challenges, emerging applications, and new developments. *Macromolecules, 56*(7), 2579–2607.
2. Brown, M. F. (2017). Soft matter in lipid–protein interactions. *Annual Review of Biophysics, 46*(1), 379–410.
3. Chaudhuri, M., Ivlev, A. V., Khrapak, S. A., Thomas, H. M., & Morfill, G. E. (2011). Complex plasma—the plasma state of soft matter. *Soft Matter, 7*(4), 1287–1298.
4. Chen, S., Tan, W. S., Bin Juhari, M. A., Shi, Q., Cheng, X. S., Chan, W. L., & Song, J. (2020). Freeform 3D printing of soft matters: Recent advances in technology for biomedical engineering. *Biomedical Engineering Letters, 10*, 453–479.
5. Christov, I. C. (2021). Soft hydraulics: From Newtonian to complex fluid flows through compliant conduits. *Journal of Physics: Condensed Matter, 34*(6), 063001.
6. Daniels, A. U., Chang, M. K., Andriano, K. P., & Heller, J. (1990). Mechanical properties of biodegradable polymers and composites proposed for internal fixation of bone. *Journal of Applied Biomaterials, 1*(1), 57–78.
7. Fuller, G. G., & Vermant, J. (2011). Dynamics and rheology of complex fluid–fluid interfaces. *Soft Matter, 7*(17), 7583–7585.
8. Grosjean, S., Wawryszyn, M., Mutlu, H., Bräse, S., Lahann, J., & Theato, P. (2019).

Soft Matter Technology at KIT: Chemical perspective from nanoarchitectures to microstructures. *Advanced Materials, 31*(26), 1806334.

9. Hatzikiriakos, S. G. (2015). Slip mechanisms in complex fluid flows. *Soft Matter, 11*(40), 7851–7856.
10. Haw, M. D. (2002). Colloidal suspensions, Brownian motion, molecular reality: A short history. *Journal of Physics: Condensed Matter, 14*(33), 7769.
11. Huber, P. (2015). Soft matter in hard confinement: Phase transition thermodynamics, structure, texture, diffusion and flow in nanoporous media. *Journal of Physics: Condensed Matter, 27*(10), 103102.
12. Kim, Y. K., Noh, J., Nayani, K., & Abbott, N. L. (2019). Soft matter from liquid crystals. *Soft Matter, 15*(35), 6913–6929.
13. Lagerwall, J. P., & Scalia, G. (2012). A new era for liquid crystal research: Applications of liquid crystals in soft matter nano-, bio- and microtechnology. *Current Applied Physics, 12*(6), 1387–1412.
14. Likos, C. N. (2006). Soft matter with soft particles. *Soft Matter, 2*(6), 478–498.
15. Lin, F., Tong, X., Wang, Y., Bao, J., & Wang, Z. M. (2015). Graphene oxide liquid crystals: Synthesis, phase transition, rheological property, and applications in optoelectronics and display. *Nanoscale Research Letters, 10*, 1–16.
16. Liu, D., & Broer, D. J. (2014). Liquid crystal polymer networks: Preparation, properties, and applications of films with patterned molecular alignment. *Langmuir, 30*(45), 13499–13509.
17. Liu, G., Zheng, H., Jiang, Z., Zhao, J., Wang, Z., Pan, B., & Xing, B. (2018). Formation and physicochemical characteristics of nano biochar: Insight into chemical and colloidal stability. *Environmental Science & Technology, 52*(18), 10369–10379.
18. Mark, J. E. (1979). Interpretation of polymer properties in terms of chain conformations and spatial configurations. *Accounts of Chemical Research, 12*(2), 49–55.
19. Nagel, S. R. (2017). Experimental soft-matter science. *Reviews of Modern Physics, 89*(2), 025002.
20. Nakayama, M., & Kato, T. (2022). Biomineral-inspired colloidal liquid crystals: From assembly of hybrids comprising inorganic nanocrystals and organic polymer components to their functionalization. *Accounts of Chemical Research, 55*(13), 1796–1808.
21. Nunes, R. W., Martin, J. R., & Johnson, J. F. (1982). Influence of molecular weight and molecular weight distribution on mechanical properties of polymers. *Polymer Engineering & Science, 22*(4), 205–228.
22. Richards, J. A., Martinez, V. A., & Arlt, J. (2021). Characterizing shear-induced dynamics in flowing complex fluids using differential dynamic microscopy. *Soft Matter, 17*(39), 8838–8849.
23. Roberto, C., & Véronique, T. (2023). Introduction to viscoelasticity and plasticity, and their relation to the underlying microscopic dynamics in soft matter systems. *Physica A: Statistical Mechanics and its Applications, 631*, 128653.

24. Saraswat, M., & Sengwa, R. J. (2024). Investigations on the optical, dielectric, electrical, and rheological properties of PEG200/ZnO semiconducting nanofluids for soft matter based technological innovations. *Hybrid Advances, 5*, 100140.
25. Stupp, S. I., Zha, R. H., Palmer, L. C., Cui, H., & Bitton, R. (2013). Self-assembly of biomolecular soft matter. *Faraday Discussions, 166*, 9–30.
26. Tabor, R. F., Chan, D. Y., Grieser, F., & Dagastine, R. R. (2011). Structural forces in soft matter systems. *The Journal of Physical Chemistry Letters, 2*(5), 434–437.
27. Tanaka, H., & Araki, T. (2006). Viscoelastic phase separation in soft matter: Numerical-simulation study on its physical mechanism. *Chemical Engineering Science, 61*(7), 2108–2141.
28. TI, Q. M., Khan, A. F., & Khan, A. (2014). Colloids in the environmental protection—Current and future trends. In *The Role of Colloidal Systems in Environmental Protection* (Vol. 4, pp. 635–677).
29. Truby, R. L., & Lewis, J. A. (2016). Printing soft matter in three dimensions. *Nature, 540*(7633), 371–378.
30. Wu, Z., Khayat, K. H., & Shi, C. (2019). Changes in rheology and mechanical properties of ultra-high-performance concrete with silica fume content. *Cement and Concrete Research, 123*, 105786.

CHAPTER

8

Modern Developments in the Physics of Matter

LEARNING OBJECTIVES

After studying this chapter, you will be able to:

1. Understand the synthesis and fabrication methods of nanomaterials;
2. Explain the unique properties of nanomaterials;
3. Describe various applications of nanomaterials in different fields;
4. Comprehend the concepts and significance of quantum materials;
5. Analyze the quantum Hall effect and its implications;
6. Identify the characteristics and applications of quantum spin liquids;
7. Explore the nature and applications of topological insulators and superconductors.

KEY TERMS FROM THIS CHAPTER

Fabrication methods
Nanotechnology
Quantum spin liquids
Topological insulators
Nanomaterials
Quantum Hall effect
Superconductors
Topological superconductors

UNIT INTRODUCTION

The field of physics of matter has witnessed significant advancements over recent decades, driven by the convergence of theoretical innovations, experimental techniques, and technological progress. These developments have expanded our understanding of materials at the atomic, molecular, and macroscopic levels, revealing new phenomena and enabling transformative applications (Arndt et al., 2012).

Nanotechnology, for instance, leverages the unique properties of materials at the nanoscale to create novel structures and devices with unprecedented functionalities. Quantum materials, which exhibit exotic properties due to quantum mechanical effects, are at the forefront of research in areas such as quantum computing and spintronics. Additionally, the discovery of topological insulators and superconductors has opened new avenues for exploring robust quantum states that are resistant to environmental disturbances, promising advances in fault-tolerant quantum computation (Cabra et al., 2012).

These modern developments are not just confined to theoretical exploration but also hold significant potential for practical applications, from next-generation electronics and energy storage solutions to medical diagnostics and treatment. The interdisciplinary nature of current research, integrating physics with materials science, chemistry, and engineering, continues to push the boundaries of what is possible, shaping the future of technology and society (Kohn, 1999).

8.1. NANOTECHNOLOGY AND NANOMATERIALS

Nanotechnology refers to the manipulation and control of matter on the atomic and molecular scale, typically below 100 nanometers. At this scale, materials exhibit unique physical and chemical properties distinct from their bulk counterparts due to quantum effects and high surface area-to-volume ratios.

For example, the electrical conductivity of materials can be tuned by altering their size, as described by the equation $\sigma = \sigma_0\left(1+\frac{a}{d}\right)$, where (σ) is the conductivity, (σ_0) is the bulk conductivity, (a) is a material-specific constant, and (d) is the diameter of the nanoparticles (Kumar et al., 2020).

Nanomaterials, which include nanoparticles, nanowires, and nanotubes, are synthesized through top-down approaches like lithography or bottom-up approaches like chemical vapor deposition. Their high surface energy makes them highly reactive, useful in catalysis, drug delivery, and electronic applications. The behavior of electrons in nanomaterials can be described using quantum confinement models, where the energy levels are quantized, and the density of states $(D(E))$ is given by:

$$D(E) = \frac{1}{2}\left(\frac{2}{2}\right)^{3/2}\sqrt{E-E}$$

This unique quantum behavior underpins the diverse applications and ongoing research in the field of nanotechnology (Kher & Kumar, 2022).

8.1.1. Synthesis and Fabrication Methods

The synthesis and fabrication of nanomaterials are critical processes in nanotechnology, enabling the creation of materials with specific properties and functions. Key methods include Chemical Vapor Deposition (CVD), the Sol-Gel process, and Electrospinning.

8.1.1.1. Chemical Vapor Deposition (CVD)

CVD is a widely used method for producing high-purity, high-performance solid materials. In this process, volatile precursors are transported in the vapor phase to a substrate where they react to form a solid film. The general reaction can be represented as:

$$\mathrm{M}_{x\mathrm{N}_y}(g) \rightarrow x\mathrm{M}(s) + y\mathrm{N}(g)$$

Remember

Remember that modern developments in the physics of matter have revolutionized our understanding and applications of materials, from nanotechnology to quantum materials and topological insulators, leading to breakthroughs in science and technology.

where (M_{xN_y}) is the precursor gas, (M) is the deposited material, and (N) is a byproduct. Parameters such as temperature, pressure, and gas flow rates are carefully controlled to achieve the desired film properties. CVD is used to deposit materials such as silicon dioxide, silicon nitride, and various metals and semiconductors (Manawi et al., 2018) (Figure 8.1).

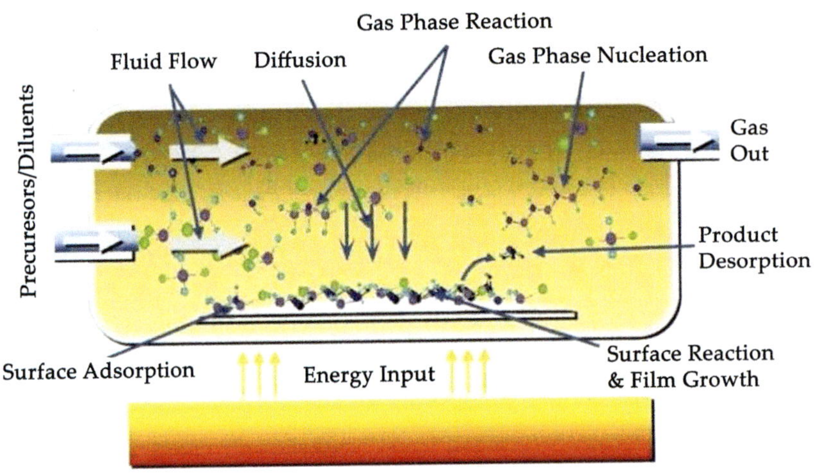

Figure 8.1. *Schematic representation of chemical vapor deposition (CVD).*

Source: Kintek, Creative Commons License.

8.1.1.2. Sol-Gel Process

The Sol-Gel process is a versatile method for fabricating ceramics and glassy materials at relatively low temperatures. It involves the transition of a system from a liquid "sol" (a colloidal suspension of particles) into a solid "gel" phase. The general steps include hydrolysis and polycondensation of metal alkoxides:

$$\mathrm{M(OR)}_n + H_2O \rightarrow \mathrm{M(OH)(OR)}_{n-1} + \mathrm{ROH}$$

$$\mathrm{M(OH)(OR)}_{n-1} \rightarrow \mathrm{M(OH)}_2\mathrm{(OR)}_{n-2} + \mathrm{ROH}$$

The gel can be further processed through drying and heating to form dense, uniform materials. Sol-Gel is employed in producing coatings, fibers, and nanoparticle-based materials (Singh et al., 2014) (Figure 8.2).

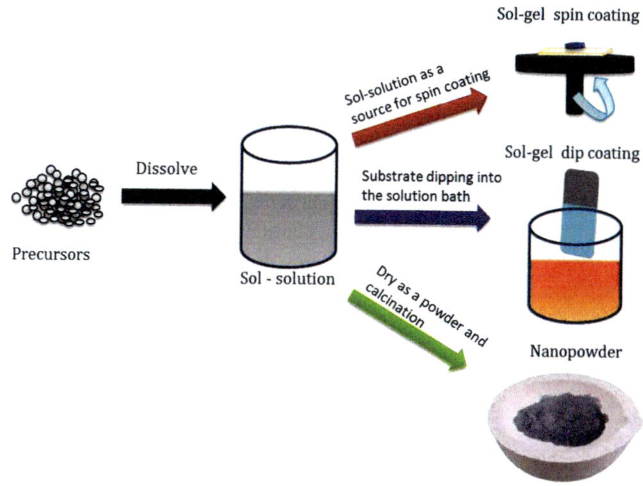

Figure 8.2. *Schematic diagram of sol-gel processing.*

Source: Shrividhya Thiagarajan, Creative Commons License.

8.1.1.3. Electrospinning

Electrospinning is a technique for creating ultrafine fibers from polymer solutions or melts using electrostatic forces. In this process, a high voltage is applied to a polymer solution, causing a jet of the solution to be ejected from the tip of a syringe needle. The jet undergoes stretching and thinning, leading to the formation of nanofibers:

$$\text{Electrostatic Force} = \frac{1}{2}\epsilon E^2$$

Where; (ϵ) is the permittivity of the medium; and (E) is the electric field. The resulting fibers are collected on a grounded substrate, creating non-woven mats with high surface area-to-volume ratios. Electrospinning is used in applications such as filtration, tissue engineering, and drug delivery systems.

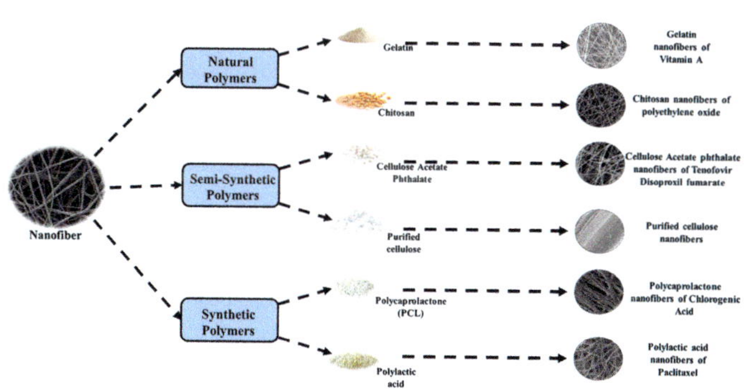

Figure 8.3. *Representation of the electrospinning method.*

Source: Abhijeet Hiwrale, Creative Commons License.

These synthesis and fabrication methods enable precise control over the structure and properties of nanomaterials, facilitating their integration into advanced technological applications (Greenfeld et al., 2011) (Figure 8.3).

8.1.2. Properties of Nanomaterials

The properties of nanomaterials are explained in the following sections.

8.1.2.1. Mechanical Properties

Nanomaterials exhibit unique mechanical properties compared to their bulk counterparts, primarily due to their reduced dimensions and high surface-to-volume ratios. The strength and hardness of nanomaterials can be significantly enhanced, as described by the Hall-Petch relationship:

$$\sigma_y = \sigma_0 + k \cdot d^{-1/2}$$

where; (σ_y) is the yield strength; (σ_0) is the friction stress; (k) is a material constant; and (d) is the grain size. For example, nanocrystalline metals can exhibit higher hardness and strength than coarse-grained metals. Additionally, the Young's modulus and fracture toughness of nanomaterials are influenced by their size and surface characteristics, making them ideal for high-strength and lightweight applications (Zhang et al., 2020).

8.1.2.2. Electrical Properties

The electrical properties of nanomaterials are governed by quantum confinement effects, which occur when the dimensions of the material are comparable to the electron's de Broglie wavelength. This leads to discrete energy levels and altered electronic band structures. For instance, the conductivity of nanomaterials can be modulated by their size and shape, as shown in the quantum confinement equation for the energy levels:

$$E_n = \frac{n^2 h^2}{8mL^2}$$

where; (E_n) is the energy level; (n) is an integer; (h) is Planck's constant; (m) is the electron mass; and (L) is the dimension of the nanomaterial. Carbon nanotubes and graphene are prime examples, exhibiting exceptional electrical conductivity and mobility, making them suitable for nanoelectronics and conductive composites (Lee et al., 2019).

8.1.2.3. Optical Properties

Nanomaterials also exhibit distinctive optical properties due to quantum size effects and surface plasmon resonance. The optical absorption and emission characteristics can be tuned by controlling the size, shape, and composition of the nanomaterials.

For instance, the energy gap (E_g) in semiconductor nanoparticles, or quantum dots, varies with particle size, influencing their photoluminescence properties:

$$E_g \approx E_{g,bulk} + \frac{2\hbar^2}{m^* R^2}$$

where; $(E_{g,bulk})$ is the bulk bandgap energy; $(\hbar)$ is the reduced Planck's constant; (m^*) is the effective mass of the electron-hole pair; and (R) is the radius of the nanoparticle. Surface plasmon resonance, observed in metal nanoparticles like gold and silver, results in strong absorption and scattering of light at specific wavelengths, leading to applications in sensing, imaging, and photothermal therapy. These unique mechanical, electrical, and optical properties of nanomaterials enable their use in a wide range of advanced technologies, from structural materials to electronic devices and biomedical applications (Ray, 2010).

8.1.3. Applications of Nanomaterials

Applications of nanomaterials are explained in the following sections.

8.1.3.1. Medicine and Drug Delivery

Nanomaterials have revolutionized medicine, particularly in drug delivery systems (Figure 8.4). Nanoparticles can be engineered to improve the bioavailability, targeted delivery, and controlled release of therapeutic agents. For instance, the diffusion coefficient (D) of nanoparticles in biological tissues can be described by the Stokes-Einstein equation.

$$D = \frac{k_B T}{6\pi\eta r}$$

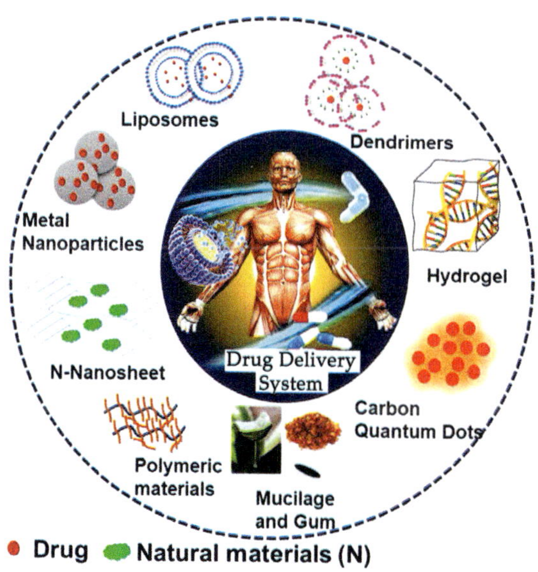

Figure 8.4. *Depiction of the different shapes and structural NMs toward drug delivery systems.*

Source: Mekala Veerapandian, Creative Commons License.

Did you know?

Did you know that nanomaterials are used in diverse applications such as drug delivery, environmental remediation, and advanced electronics due to their unique size-dependent properties? These applications highlight the versatility and potential impact of nanotechnology.

where; (k_B) is the Boltzmann constant; (T) is the temperature; (η) is the viscosity of the medium; and (r) is the radius of the nanoparticle. This allows for precise control over the distribution and release rate of drugs at specific sites, reducing side effects and improving therapeutic efficacy. Additionally, nanomaterials like liposomes, dendrimers, and polymeric nanoparticles are used in targeted cancer therapy, imaging, and diagnostics (Joshi et al., 2019).

8.1.3.2. Environmental Applications

Nanomaterials play a crucial role in environmental applications, particularly in water treatment and pollution remediation. Titanium dioxide (TiO2) nanoparticles, for example, are used in photocatalytic degradation of pollutants. The photocatalytic activity is governed by the generation of electron-hole pairs under UV light, which can be represented by:

$$TiO_2 + h\nu \rightarrow TiO_2\left(e^- + h^+\right)$$

The generated electrons and holes react with water and oxygen to form reactive oxygen species (ROS), which degrade organic pollutants. Nanomaterials are also used in developing advanced filtration membranes for water purification, removing contaminants such as heavy metals, pathogens, and organic compounds, thereby enhancing the quality of drinking water and reducing environmental pollution (Mazari et al., 2021).

8.1.3.3. Electronics and Photonics

In electronics and photonics, nanomaterials have enabled the development of smaller, faster, and more efficient devices. Quantum dots, for instance, are used in light-emitting diodes (LEDs), displays, and solar cells due to their tunable optical properties. The efficiency of quantum dot solar cells can be enhanced by the multiple exciton generation (MEG) process, where a single high-energy photon generates multiple electron-hole pairs:

$$\eta_{\text{MEG}} = \frac{E_{\text{photon}}}{E_{\text{bandgap}}}$$

where; (η_{MEG}) is the MEG efficiency; (E_{photon}) is the energy of the incident photon; and (E_{bandgap}) is the bandgap energy of the quantum dot. Carbon nanotubes and graphene are also used in transistors, sensors, and flexible electronics due to their exceptional electrical conductivity, mechanical strength, and thermal stability. These applications demonstrate the transformative impact of nanomaterials on the electronics and photonics industries, leading to more advanced and sustainable technologies.

In summary, the unique properties of nanomaterials facilitate their integration into diverse applications, driving innovations in medicine, environmental protection, and electronic devices (Diao & Wang, 2018).

8.2. QUANTUM MATERIALS

Quantum materials exhibit unique properties governed by quantum mechanical effects that emerge at macroscopic scales. These materials include superconductors, topological insulators, and quantum spin liquids, characterized by phenomena such as electron pairing, non-trivial topological order, and entanglement. The behavior of quantum materials is described by the Schrödinger equation: $H^{\psi} = E\psi \; where \hat{H}$ is the Hamiltonian operator, ψ is the wave function, and EEE is the energy eigenvalue. Quantum materials have profound implications for advanced technologies, including quantum computing, spintronics, and low-power electronics, driving significant advancements in condensed matter physics and materials science (Bauer et al., 2020).

8.2.1. Quantum Hall Effect

The integer quantum Hall effect (IQHE) occurs in two-dimensional electron systems subjected to low temperatures and strong magnetic fields, where the Hall conductance (σ_{xy}) quantizes in integer multiples of (e^2/h):

$$\sigma_{xy} = \frac{ne^2}{h}$$

where; (e) is the electron charge; (h) is Planck's constant; and $(\ )$ is a non-negative integer. The quantization arises from the formation of discrete Landau levels, where electrons occupy cyclotron orbits with quantized energy levels:

$$E_n = \left(n + \frac{1}{2}\right)\hbar\omega_c$$

where; (n) is a non-negative integer; $(\hbar)$ is the reduced Planck's constant; and $(\omega_c = eB/m^*)$ is the cyclotron frequency with (B) as the magnetic field strength and (m^*) as the effective mass of the electron. The IQHE demonstrates robustness against impurities and defects, making it a key phenomenon in precision measurements of fundamental constants (Figure 8.5).

The fractional quantum Hall effect (FQHE) occurs under similar conditions but at even higher magnetic fields and lower temperatures, where the Hall conductance quantizes in fractional multiples of

$\left(e^2 / h\right)$:

$$\sigma_{xy} = \frac{p}{q}\frac{e^2}{h}$$

with (p) and (q) being integers. The FQHE is explained by the formation of composite fermions, where electrons bind with an even number of magnetic flux quanta to form new quasi-particles (Stormer, 1999).

Tip

The Quantum Hall Effect, observed in two-dimensional electron systems under low temperatures and high magnetic fields, provides a precise method for measuring fundamental constants and has applications in metrology and quantum computing.

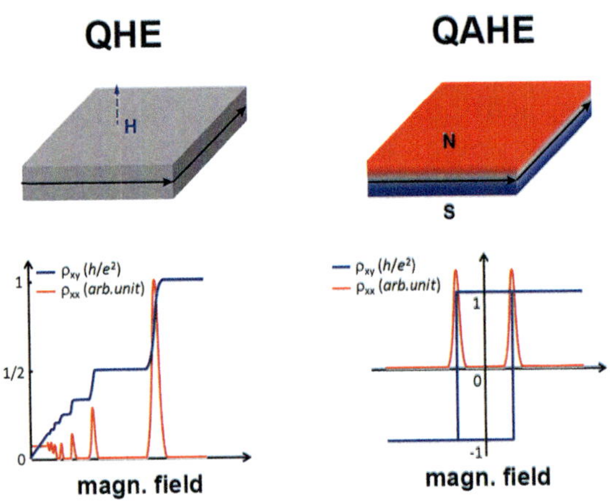

Figure 8.5. *(Left) The quantum Hall effect (QHE) occurs in a two-dimensional electron system under a strong magnetic field. (Right) The quantum anomalous Hall effect shows quantum Hall-like behavior without an applied magnetic field.*

Source: A. J. Bestwick, Creative Commons License.

The Laughlin wave function describes the ground state of these systems:

$$\psi\left(z_1, z_2, \ldots, z_N\right) = \prod_{i<j}\left(z_i - z_j\right)^m e^{-\sum_k |z_k|^2 / 4l_B^2}$$

where; $\left(z_k = x_k + iy_k\right)$ represents the complex coordinates of the k-th electron; (m) is an odd integer; and $\left(l_B = \sqrt{\hbar / eB}\right)$ is the magnetic length. The FQHE reveals the rich interplay of electron interactions and topology, leading to new states of matter with potential applications in quantum computing (Maciejko et al., 2011).

8.2.2. Quantum Spin Liquids

Quantum spin liquids (QSLs) are exotic states of matter where magnetic moments remain disordered even at absolute zero temperature due to quantum fluctuations. Theoretical models of QSLs include the Kitaev model and the Resonating Valence Bond (RVB) theory. The Kitaev model describes spins on a honeycomb lattice with bond-dependent interactions:

$$H = -J_x \sum_{ij_x} \sigma_i^x \sigma_j^x - J_y \sum_{ij_y} \sigma_i^y \sigma_j^y - J_z \sum_{ij_z} \sigma_i^z \sigma_j^z$$

where; $\left(\sigma_i^\alpha\right)$ are Pauli matrices; and $\left(J_x, J_y, J_z\right)$ are coupling constants. The RVB theory, proposed by Anderson, describes QSLs as superpositions of singlet pairs, with the wave function:

$$|\psi_{\mathrm{RVB}} = \sum_{\text{dimer coverings}} a_c |c$$

where; $\left(a_c\right)$ are coefficients; and $\left(|c\right)$ represents different dimer configurations (Lee, 2007). Experimental realizations of QSLs have been observed in materials like Herbertsmithite, $\left(\mathrm{ZnCu_3(OH)_{6Cl_2}}\right)$, where no magnetic ordering is detected down to millikelvin temperatures. Neutron scattering experiments reveal a continuum of excitations, indicative of fractionalized excitations, such as spinons, in QSLs. Specific heat measurements at low temperatures show deviations from conventional magnetic behavior, suggesting the presence of gapless excitations. The kagome lattice structure in Herbertsmithite provides a geometrically frustrated environment, preventing spins from aligning and promoting the formation of a QSL. Other candidates include $\left(\mathrm{EtMe_{3Sb}[Pd(dmit)_2]_2}\right)$ and $\left(\mathrm{YbMgGaO_4}\right)$, which also exhibit properties consistent with QSL behavior. These discoveries open pathways to exploring new quantum phases and potential applications in quantum information and materials science (Gingras & McClarty, 2014).

8.2.3. Applications of Quantum Materials

Applications of quantum materials are explained in the following sections.

8.2.3.1. Quantum Computing

Quantum materials play a crucial role in the development of quantum computing. Superconductors, such as niobium and aluminum, are used to create qubits, the fundamental units of quantum information. In superconducting qubits, information is stored in the quantized energy levels of Cooper pairs, and the Josephson junction facilitates quantum coherence and tunneling. The Hamiltonian for a superconducting qubit can be expressed as:

$$H = \frac{Q^2}{2C} - E_J \cos(\phi)$$

where; (Q) is the charge; (C) is the capacitance; $\left(E_J\right)$ is the Josephson energy; and (ϕ) is the phase difference across the junction. Quantum error correction and entanglement are essential for building scalable quantum computers, and materials with low decoherence rates are crucial for maintaining qubit fidelity (Lordi & Nichol, 2021).

8.2.3.2. Quantum Sensing

Quantum materials enhance the sensitivity and precision of sensors beyond classical limits. Nitrogen-vacancy (NV) centers in diamond are prominent examples, utilized for detecting magnetic and electric fields at the nanoscale. The NV center's ground state is a spin triplet, and its Hamiltonian is given by:

$$H = DS_z^2 + E\left(S_x^2 - S_y^2\right) + \mu_B \vec{B} \cdot \vec{S}$$

where; (D) is the zero-field splitting; (E) is the strain-induced splitting; (μ_B) is the Bohr magneton; and $(\vec{S})$ represents the spin-1 operators. Quantum sensing applications extend to biological imaging, navigation, and materials characterization, exploiting the quantum coherence of materials to achieve unparalleled sensitivity. These advancements leverage the unique properties of quantum materials to revolutionize technologies in precision measurement and imaging (Raparthi, 2022).

8.3. TOPOLOGICAL INSULATORS AND SUPERCONDUCTORS

Topological insulators are quantum materials that exhibit conducting surface states protected by time-reversal symmetry. The hallmark of topological insulators is the presence of a bulk band gap and metallic surface states with Dirac-like dispersion relations. The Hamiltonian for a time-reversal invariant topological insulator can be described by (Hasan et al., 2015): $(\tau_z)(M(\boldsymbol{k}))(A_i(\boldsymbol{k}))$ are momentum-dependent parameters, and (σ_i) are Pauli matrices. The surface states are protected by the topology of the bulk wavefunctions, leading to robust conductance against disorder and imperfections.

Prominent examples include $(\mathrm{Bi}_{2\mathrm{Se}_3})$ and $(\mathrm{Bi}_{2\mathrm{Te}_3})$, which exhibit quantized conductance at the surface due to their non-trivial topology (Chiu et al., 2013).

8.3.1. Topological Superconductors

Topological superconductors are a class of materials that combine superconductivity with non-trivial topology, hosting Majorana fermions as emergent quasiparticles. The Hamiltonian for a topological superconductor typically includes terms that break time-reversal symmetry and induce pairing between electrons with opposite momenta and spins. A notable example is the $(\mathrm{d}_{x^2-y^2} + i\mathrm{d}_{xy})$ pairing state, which can support Majorana bound states at interfaces or defects. The topological properties are characterized by topological invariants such as the Chern number or Z2 index, reflecting the non-trivial topology of the wavefunction (Xu et al., 2016).

8.3.2. Experimental Techniques and Discoveries

Experimental techniques for studying topological materials include angle-resolved photoemission spectroscopy (ARPES), scanning tunneling microscopy (STM), and quantum transport measurements. ARPES provides direct information about the band structure and Fermi surface topology, crucial for identifying topological surface states. STM enables the visualization of atomic-scale details and local density of states, revealing the spatial distribution of surface states and potential defects. Quantum transport measurements, such as Hall conductance and magneto-resistance, confirm the quantized conductance and topological protection of surface states in topological insulators (Leijnse & Flensberg, 2012).

Recent discoveries include the observation of edge states in topological insulators and the realization of Majorana bound states in nanowires coupled to superconductors, demonstrated by zero-bias conductance peaks in tunneling spectroscopy. These findings underscore the potential of topological materials for novel quantum technologies and fundamental physics research (Sharma et al., 2022).

8.3.3. Applications of Topological Materials

Applications of topological materials span quantum computing, spintronics, and quantum information technologies. Topological insulators are promising candidates for spintronic devices due to their spin-momentum locking and low dissipation properties. Quantum computing benefits from Majorana fermions in topological superconductors for robust qubit implementations and fault-tolerant quantum information processing. Majorana bound states are crucial for topological quantum computation, offering protection against decoherence (Li & Wei, 2021).

Moreover, topological materials show potential in creating efficient thermoelectric devices, exploiting the unique electronic properties of topological surface states for enhanced thermoelectric efficiency. Advances in material synthesis and characterization techniques continue to drive the exploration of new topological phases and their applications in next-generation electronics and quantum technologies (Tian et al., 2017).

ACTIVITY 8.1.

Explore experimental techniques used to study topological insulators and superconductors. Conduct hands-on demonstrations or simulations to understand how these materials exhibit unique electronic properties critical for future technologies.

SUMMARY

- Chapter Eight, titled "Modern Developments in the Physics of Matter," delves into the forefront of contemporary research and applications in material science. It begins with an exploration of nanotechnology and nanomaterials, detailing various synthesis and fabrication methods such as chemical vapor deposition, sol-gel processes, and electrospinning. These methods enable the creation of nanomaterials with unique properties, such as high surface area to volume ratio, which distinguishes them from their bulk counterparts. The chapter highlights the diverse applications of nanomaterials in fields such as medicine, electronics, and environmental science.
- The chapter then transitions to quantum materials, focusing on phenomena such as the quantum Hall effect and quantum spin liquids. The quantum Hall effect, observed in two-dimensional electron systems under low temperatures and strong magnetic fields, is significant for its precision in measuring fundamental constants. Quantum spin liquids, characterized by the absence of magnetic order even at absolute zero, present intriguing possibilities for future technologies due to their exotic properties.
- Topological insulators and superconductors form the final section of the chapter. Topological insulators are materials that conduct electricity on their surfaces while remaining insulating in their bulk. These materials have potential applications in spintronics and quantum computing. The chapter also discusses topological superconductors, which are a focus of intense research due to their potential to host Majorana fermions, particles that could revolutionize quantum computation.

REVIEW QUESTIONS

1. What are the primary methods used for the synthesis and fabrication of nanomaterials?
2. Explain the unique properties that distinguish nanomaterials from bulk materials.
3. Discuss at least three applications of nanomaterials.
4. What is the quantum Hall effect, and why is it significant in modern physics?
5. Describe the concept of quantum spin liquids and their potential applications.
6. How do topological insulators differ from conventional insulators, and what are their applications?

MULTIPLE CHOICE QUESTIONS

1. Which of the following is NOT a method for synthesizing nanomaterials?

a. Chemical Vapor Deposition

b. Sol-Gel Process

c. Electrospinning

d. Classical Mechanics

2. What is a key property of nanomaterials that differentiates them from bulk materials?

a. Electrical Conductivity

b. Large Surface Area to Volume Ratio

c. Thermal Expansion

d. Elasticity

3. The quantum Hall effect is primarily observed in which type of material?

a. Superconductors

b. Insulators

c. Two-Dimensional Electron Systems

d. Metals

4. Quantum spin liquids are characterized by:

a. Ordered magnetic moments

b. Absence of magnetic order at low temperatures

c. High electrical conductivity

d. High thermal conductivity

5. Topological insulators are known for:

a. Conducting electricity on the surface but acting as insulators in the bulk

b. High thermal conductivity throughout the material

c. Superconducting properties at room temperature

d. Magnetic order at low temperatures

Answers to Multiple Questions

1. (d); 2. (b); 3. (c); 4. (b); 5. (a).

REFERENCES

1. Arndt, M., Ekers, A., Von Klitzing, W., & Ulbricht, H. (2012). Focus on modern frontiers of matter wave optics and interferometry. *New Journal of Physics, 14*(12), 125006.
2. Bauer, B., Bravyi, S., Motta, M., & Chan, G. K. L. (2020). Quantum algorithms for quantum chemistry and quantum materials science. *Chemical Reviews, 120*(22), 12685–12717.
3. Cabra, D. C., Honecker, A., & Pujol, P. (2012). *Modern Theories of Many-Particle Systems in Condensed Matter Physics* (Vol. 843, pp. 87–90). Springer Science & Business Media.

4. Chiu, C. K., Yao, H., & Ryu, S. (2013). Classification of topological insulators and superconductors in the presence of reflection symmetry. *Physical Review B—Condensed Matter and Materials Physics, 88*(7), 075142.
5. Diao, F., & Wang, Y. (2018). Transition metal oxide nanostructures: Premeditated fabrication and applications in electronic and photonic devices. *Journal of Materials Science, 53*, 4334–4359.
6. Gingras, M. J., & McClarty, P. A. (2014). Quantum spin ice: A search for gapless quantum spin liquids in pyrochlore magnets. *Reports on Progress in Physics, 77*(5), 056501.
7. Greenfeld, I., Arinstein, A., Fezzaa, K., Rafailovich, M. H., & Zussman, E. (2011). Polymer dynamics in semidilute solution during electrospinning: A simple model and experimental observations. *Physical Review E—Statistical, Nonlinear, and Soft Matter Physics, 84*(4), 041806.
8. Hasan, M. Z., Xu, S. Y., & Bian, G. (2015). Topological insulators, topological superconductors and Weyl fermion semimetals: Discoveries, perspectives and outlooks. *Physica Scripta, 2015*(T164), 014001.
9. Joshi, K., Mazumder, B., Chattopadhyay, P., Bora, N. S., Goyary, D., & Karmakar, S. (2019). Graphene family of nanomaterials: Reviewing advanced applications in drug delivery and medicine. *Current Drug Delivery, 16*(3), 195–214.
10. Kher, C., & Kumar, S. (2022). The application of nanotechnology and nanomaterials in cancer diagnosis and treatment: A review. *Cureus, 14*(9), 987–990.
11. Kohn, W. (1999). An essay on condensed matter physics in the twentieth century. *Reviews of Modern Physics, 71*(2), 59.
12. Kumar, R., Aadil, K. R., Ranjan, S., & Kumar, V. B. (2020). Advances in nanotechnology and nanomaterials-based strategies for neural tissue engineering. *Journal of Drug Delivery Science and Technology, 57*, 101617.
13. Lee, H. K., Nam, I. W., Tafesse, M., & Kim, H. K. (2019). Fluctuation of electrical properties of carbon-based nanomaterials/cement composites: Case studies and parametric modeling. *Cement and Concrete Composites, 102*, 55–70.
14. Lee, P. A. (2007). From high temperature superconductivity to quantum spin liquid: Progress in strong correlation physics. *Reports on Progress in Physics, 71*(1), 012501.
15. Leijnse, M., & Flensberg, K. (2012). Introduction to topological superconductivity and Majorana fermions. *Semiconductor Science and Technology, 27*(12), 124003.
16. Li, Z., & Wei, B. (2021). Topological materials and topologically engineered materials: Properties, synthesis, and applications for energy conversion and storage. *Journal of Materials Chemistry A, 9*(3), 1297–1313.
17. Lordi, V., & Nichol, J. M. (2021). Advances and opportunities in materials science for scalable quantum computing. *MRS Bulletin, 46*, 589–595.
18. Maciejko, J., Hughes, T. L., & Zhang, S. C. (2011). The quantum spin Hall effect. *Annual Review of Condensed Matter Physics, 2*(1), 31–53.

19. Manawi, Y. M., Ihsanullah, Samara, A., Al-Ansari, T., & Atieh, M. A. (2018). A review of carbon nanomaterials' synthesis via the chemical vapor deposition (CVD) method. *Materials, 11*(5), 822.

20. Mazari, S. A., Ali, E., Abro, R., Khan, F. S. A., Ahmed, I., Ahmed, M., & Shah, A. (2021). Nanomaterials: Applications, waste-handling, environmental toxicities, and future challenges—A review. *Journal of Environmental Chemical Engineering, 9*(2), 105028.

21. Raparthi, M. (2022). Quantum sensing technologies for biomedical applications: Investigating the advancements and challenges. *Journal of Computational Intelligence and Robotics, 2*(1), 21–32.

22. Ray, P. C. (2010). Size and shape dependent second order nonlinear optical properties of nanomaterials and their application in biological and chemical sensing. *Chemical Reviews, 110*(9), 5332–5365.

23. Sharma, M. M., Sharma, P., Karn, N. K., & Awana, V. P. S. (2022). Comprehensive review on topological superconducting materials and interfaces. *Superconductor Science and Technology, 35*(8), 83003.

24. Singh, L. P., Bhattacharyya, S. K., Kumar, R., Mishra, G., Sharma, U., Singh, G., & Ahalawat, S. (2014). Sol-Gel processing of silica nanoparticles and their applications. *Advances in Colloid and Interface Science, 214*, 17–37.

25. Stormer, H. L. (1999). Nobel lecture: The fractional quantum Hall effect. *Reviews of Modern Physics, 71*(4), 875.

26. Tian, W., Yu, W., Shi, J., & Wang, Y. (2017). The property, preparation and application of topological insulators: A review. *Materials, 10*(7), 814.

27. Xu, G., Lian, B., Tang, P., Qi, X. L., & Zhang, S. C. (2016). Topological superconductivity on the surface of Fe-based superconductors. *Physical Review Letters, 117*(4), 47001.

28. Zhang, J., Liu, Y., Cui, L., Hao, S., Jiang, D., Yu, K., & Yang, H. (2020). "Lattice strain matching" -enabled nanocomposite design to harness the exceptional mechanical properties of nanomaterials in bulk forms. *Advanced Materials, 32*(18), 1904387.

Index

C

D

T

U

Printed and bound by CPI Group (UK) Ltd, Croydon, CR0 4YY
18/09/2026
14965070-0001